LOGIC, LAWS, AND LIFE

Volume 6 *University of Pittsburgh Series in the Philosophy of Science*

Logic,

Editor

ROBERT G. COLODNY

LEONARD J. SAVAGE

RONALD N. GIERE

ROBERT E. BUTTS

MICHAEL RUSE

PETER MACHAMER

KENNETH F. SCHAFFNER

HADLEY CANTRIL

ABNER SHIMONY

ROBERT EFRON

MAURICE MANDELBAUM

Laws, & Life

Some Philosophical Complications

University of Pittsburgh Press

Published by the University of Pittsburgh Press, Pittsburgh, Pa. 15260

Feffer and Simons, Inc., London
Manufactured in the United States of America

Library of Congress Cataloging in Publication Data

Main entry under title:

Logic, laws, and life.

(University of Pittsburgh series in the philosophy of science: v. 6)
Essays which have grown out of lectures presented at the annual lecture series conducted by the Center for Philosophy of Science at the University of Pittsburgh.
Includes indexes.
CONTENTS: Colodny, R. G. Introduction.—Savage, L. J. The shifting foundations of statistics.—Butts, R. E. Consilience of inductions and the problem of conceptual change in science.—Ruse, M. Is biology different from physics? [etc.]
1. Life sciences—Philosophy—Addresses, essays. I. Savage, Leonard J. II. Colodny, Robert Garland. III. Series: Pittsburgh. University. University of Pittsburgh series in the philosophy of science; v. 6.
QH331.L63 574′.01 76-50886
ISBN 0-8229-3346-2

Contents

Preface

Each year, the Center for Philosophy of Science at the University of Pittsburgh, with the cosponsorship of the Department of History and Philosophy of Science, invites about ten philosophers or scientists to present a public lecture on some topic in the philosophy of the physical, biological, or social sciences.

The essays in the present volume, like those in some of the earlier volumes of the University of Pittsburgh Series in the Philosophy of Science, have grown out of such lectures, which the Center has conducted since 1960.

I thank Nicholas Rescher, Laurens Laudan, W. W. Bartley, III, Kenneth Schaffner, and Allen Janis for having helped me at one time or another during this period with the planning or administration of this program by serving as Associate Director.

This volume is dedicated to the memory of Professor Leonard J. Savage, whose contribution to it is based on a lecture he delivered in our program during the last year of his fruitful life.

ADOLF GRÜNBAUM
Andrew Mellon Professor of Philosophy and
Director of the Center for Philosophy of Science,
University of Pittsburgh

ROBERT G. COLODNY
University of Pittsburgh

Introduction

> Within the next few decades discoveries in biology can radically change human life as we know it. While it is true that biology has already had a substantial impact on medicine and food production, by comparison to future possibilities these effects are piecemeal and trivial. Advances in the treatment of disease, in understanding the processes of ageing, in the control of behaviour, in the harnessing of micro-organisms to produce particular chemicals, and in food production, have potentialities which will expose the limitations of the social and political structures which have evolved for the application of scientific and technical advances. The problem is not simply that we may not derive the maximum benefit from these advances but that if this new information is not correctly applied, our well-being or even the survival of our species is threatened. We are threatened physically as a species by the depletion of resources and poisoning of our environment because of unplanned technological developments. We are threatened socially as a species by the techniques for the control of behaviour which could dehumanize and could destroy creativity. As individuals, our value systems are being undermined as new knowledge on biological phenomena such as heredity and behaviour undermines the myths and dogmas at the heart of our ethical and religious beliefs.
>
> —Watson Fuller
> *The Biological Revolution, Social Good or Social Evil?*

> By adding to the knowledge of man's biological nature, science helps the humanist better to understand the human condition, and to define the good life.
>
> —René Dubos
> *"Humanistic Biology"*

When volume 1 of this series, *Frontiers of Science and Philosophy,* was published in 1962, the noted geneticist Ernst Caspari noted the following: "If one reads the recent literature on the Philosophy of Science, one is struck by the fact that it deals almost exclusively with the inorganic sciences, particularly physics. The reason seems to be that physics is the oldest and most basic of the modern natural sciences, and has as a consequence, developed further than others." Caspari also noted that the life sciences in recent times have not been extensively studied as a separate subject of philosophical inquiry. The editor of this series attended the International Conference of Cellular Biology held at Brown University in 1962, having gone there in search of a philosophically "oriented" biologist. Five days of intensive discussions with leading members of that distinguished conclave failed to turn up a single scholar who would confess to philosophical interests or who thought that philosophical analysis of biological concepts or methods was of any pressing relevance. This was odd for a discipline whose founder, Aristotle, had been a philosopher of some eminence. As Caspari went on to remark, biologists at the turn of this century had engaged in vigorous controversy about the philosophical basis of their science, but the rapid advance of biological knowledge had endowed those disputes with only an antiquarian value.

Undoubtedly the rapid advance of biology in roughly the span of the last generation constitutes one of the most fascinating and startling episodes in the intellectual history of our species. It also appears to be the case that this biological revolution has been stimulated and sustained by the import into the life sciences generally of ideas and instruments developed by neighboring disciplines such as physics, chemistry, and mathematics. Furthermore, experts from these "older" fields have themselves become workers in specialized fields of biology such as biochemistry and biophysics. That somewhat notorious success story associated with DNA and the double helix is but one celebrated event among many more pedestrian reports that are lost in the daily floods of research papers.

This aspect of the syncretic process which is so freighted with philosophical implications and complications is strangely reminiscent of an epoch in the history of nineteenth-century physics, when that supposedly most mature discipline was catalyzed by imports from the community of biologists: the revival of the wave theory of light by the physician Thomas Young; the postulation of the first law of thermodynamics by the physician Julius Mayer; and the contributions to various fields of physics by the physician Hermann Ludwig Helmholtz.

To give a sense of the rapidity with which biology has been transformed, consider these four mementos from the history of biology:

There are periods in the history of every science when a false hypothesis is better than none at all, but is a necessary forerunner of, and preparation for, the true one. (T. H. Huxley, "The Cell Theory," 1853)

But the zoologist or morphologist has been slow, where the physiologist has long been eager, to invoke the aid of the physical or mathematical sciences; and the reasons for this difference lie deep, and are partly rooted in old tradition and partly in the diverse minds and temperaments of men. To treat the living body as a mechanism was repugnant, and seemed even ludicrous; Pascal and Goethe, lover of nature as he was, ruled mathematics out of place in natural history. Even now the zoologist has scarce begun to dream of defining in mathematical language even the simplest organic forms. When he meets with a simple geometrical construction, for instance in the honeycomb, he would fain refer it to psychical instinct, or to skill and ingenuity, rather than to the operation of physical forces or mathematical laws; when he sees in snail, or nautilus, or tiny foraminiferal or radiolarian shell a close approach to sphere or spiral, he is prone of old habit to believe that after all it is something more than a spiral or a sphere, and that in this "something more" there lies what neither mathematics nor physics can explain. In short, he is deeply reluctant to compare the living with the dead, or to explain by geometry or by mechanics the things which have their part in the mystery of life. (D'Arcy Wentworth Thompson, *On Growth and Form,* 1917)

The story of genetics, and especially of the gene concept, is one of the most remarkable chapters in the history of the scientific enterprise. From Gregor Mendel and his garden peas to the American zoologist Thomas Hunt Morgan (1866–1945) and his fruit flies to the contemporary biologists James D. Watson and Francis Crick and the double helix of DNA, geneticists have unraveled the surprisingly simple rules that underlie the superficially bewildering complexities of heredity. They have interpreted those rules in terms of specific material objects, the genes. They have converted genetics from a purely biological science to a chemical one—molecular genetics—and have identified the individual genes with specific portions of nucleic acid molecules. They have deciphered the chemical script in which the instructions for the function of the genes are embodied, the devices by which the genes are copied when new cells are made, and the decoding apparatus that translates the chemical script of the genes into the chemical structure of proteins, which are the key products of the genes. Today man looks upon the specific materials of heredity, including his own, from the vantage point of a comprehensive, intellectually satisfying framework of knowledge. Future research will undoubtedly add new findings, but the basic structure of biology, resting on the twin foundations of evolution theory and molecular genetics, is here to stay, just as the basic framework of physics, resting on atomic theory, quantum mechanics, and relativity, is unshakably firm. (S. E. Luria, *Life—The Unfinished Experiment,* 1974)

We can split the atom, we can safely visit the moon, but of the mechanisms of differentiation we know next to nothing and of our cerebral processes . . . of this we know nothing. Let us put our knowledge of differentiation in perspective with the splitting of the atom: we are probably at the stage where the English philosopher Dalton defined the atom in 1805. As far as our knowledge of mental function goes, we are in the pre-Copernican era of cosmology. (Ernest Borek, *The Sculpture of Life,* 1973)

This is not the first time that the bitter paradox of human groping for understanding of the natural world has been revealed. Knowledge illuminates vast vistas of ignorance.

But what is more significant are the style and tempo of advance in the biological disciplines—not only the exquisite techniques, the refined instrumentation, the international cooperation, but the philosophical sophistication of the best work and its easy symbiotic relations with all other disciplines. That one can discover petty orthodoxies, faddish cliques, and imperial pretensions here and there merely testifies to the human component in this most important of contemporary human dramas.

It is clear that for better or for worse, the advances in biology not only will remake our image of the natural world but will mandate drastic changes in our social arrangements and patterns of political possibilities. Here all human activities come together, as is suggested by such terms as sociobiology, bioeconomics, exobiology, bioethics, and so forth.

A far too large percentage of the population tended to regard the profound revolutions in physics as a kind of spectator sport until mushroom clouds over Hiroshima and Nagasaki and radioactive milk indicated that physics *was* the public's business. Biology, with its genetic engineering, behavior modification, its relation to food supply, population control, organ transplants, general ecology, and above all with its implicit capacity to define what is human, ought to be Everyman's concern. The claims to certainty or probability asserted by its practitioners fall in the province of philosophy. This volume makes a very modest effort to survey some of this terrain.

"In principle," writes L. J. Savage, "statistics is applicable in every area in which there is systematic empirical study." Because the biological disciplines are profoundly engaged in studies of this type, it is clear that statistics is one of their vital tools of analysis; and it is equally evident that the *status* of biological theories and concepts will reflect the varied theories of statistical inference as well as the shifting ideas about inductive reasoning. "How to behave in the face of uncertainty" is one of the justifications Savage gives for concern with the foundations of statistics. As Savage relates, statisticians were consulted on the design of the experiment which tested the Salk vaccine; and this was an issue involving the health of millions. It is apparent that the prosperity and happiness of other millions are affected by statistical analyses that deal with economic and social life in the modern world.

R. N. Giere, who advocates a view of statistical inference different from that of Savage, develops his theory in subtle detail and demonstrates its practical advantages in the analysis of certain concrete problems of Mendelian genetics. Although the example is chosen for heuristic reasons, both the practical implications and the deep theoretical import of these arguments emerge with great clarity.

The history of science and of its attendant philosophies is interlaced with an unbroken sequence of arguments derived from the problems of inductive reasoning. From Hume and Whewell to Carnap and Salmon, these debates have focused on the logical issues of confirmability and certainty and have illuminated our understanding of the processes of conceptual change in science. Robert E. Butts, in his reexamination of this well-mined vein of philosophy, has indeed revealed the new and permanent value of such studies—particularly in reference to the biological disciplines, with their reliance on long chains of inductive consiliences.

Michael Ruse poses the question: "Is biology different from physics?" not to raise ancient issues of vitalism versus mechanism, but to examine assertions that biology has a set of unique characteristics which separate it from physics as a mode of scientific procedures and which have different standards of conceptual rigor. A careful analysis of the theoretical structure of population genetics—a basic pillar of contemporary biological thought—reveals deep formal affinities with statistical mechanics. Furthermore, a close examination of teleological argumentation uncovers no reasons to believe that a rigorous causal order will not eventually emerge as the foundation of evolutionary theory. As Ruse concludes: "In short, all other things being equal, biology is like physics—but not quite everything is equal."

Peter Machamer examines the logical structure of teleological explanation, particularly with reference to the use of this mode of analysis in theories of natural selection. He concludes *contra* Ruse that function-adaptation processes over time are probably not reducible to "finer-grain explanations."

Practically all of the philosophical issues inventoried in the preceding essays are brought together in Kenneth Schaffner's study. Here the scientific data examined are located at the frontier of biological knowledge, the beautiful work of the molecular biologists as it pertains to the relation of DNA-enzymes-protein synthesis. The novelty of Schaffner's essay is not to be found in the analysis of the reductionist program per se but rather in the analysis of the value judgments that are an irreducible ingredient of the scientist's theory. His conclusions stand in sharp contrast to certain traditional positivistic models.

Finally we come to the group of papers by Hadley Cantril, Abner Shimony, Robert Efron, and Maurice Mandelbaum. Here the concerns center on problems of psychology—in precisely that disputed no man's land of earlier times where biology and philosophy overlap. Cantril examines various aspects of psychological research which have as their goal that kind of knowledge which (if universalized) would permit "individuals to live more effectively." He ranges over received views of scientific and philosophical ideas that span the time from Polybius and Plutarch to Pavlov and contemporary behaviorists. He concludes: "The major advances in science are due much more to improved formulations and the discovery of new variables than they are to improved or new methodologies, although the latter must, of course, never be underrated. The scientific inquiry that we call psychology will increase our understanding of experience and behavior insofar as it can differentiate the processes that play a role, demonstrate how the variations of certain processes affect other processes, and how all of them together transact to constitute the orchestration of biological, personal, and social living."

The extent to which that austere branch of philosophy, epistemology, may be enriched by findings of empirical psychology is the main concern of Abner Shimony. Here he engages in a critical dissection of some ideas advanced by the late Norwood R. Hanson, who had made strong claims for the Kantian proposition that the very process of perception was affected by antecedent conceptual commitment. This is a many-faceted argument, and its conclusions suggest research programs which would have as their goal a stronger linkage between advancing psychological and biological sciences and the older academic philosophies of mind.

Aspects of this last problem are considered by Robert Efron in his searching critique of the reductionist methodology as it impinges on the understanding and the scientific status of "consciousness."

Dr. Efron begins with the assertion that "the science of biology suffers from a progressive and potentially fatal epistemological disorder. It is characterized by such profound chaos in the realm of definitions and the logical relationships between concepts that many biologists appear to have lost cognitive contact with reality. One of the most fundamental causes of this disorder is a philosophical principle: It holds that all the phenomena of life will ultimately be reduced to, that is, accounted for, described by, and deduced from, the laws of physics and chemistry. It is known as the principle of reduction."

Efron argues that this program has been sustained for the past century largely by semantic legerdemain and logical error. He traces this process through the history of the concepts of reflex action and memory. Despite his pessimistic overview of the relations now obtaining between biology and some realms of philosophy, his essay is itself evidence for the proposition that the unification of discrete areas of scientific inquiry need not lead permanently to the compounding of their separate errors.

Finally, Mandelbaum surveys the relationship between psychology and sociology. Here are combined themes previously sounded by Cantril, Shimony, and Efron. Mandelbaum shows that a dialectical understanding of the interaction of the principles of the two disciplines must be maintained if one is to "explain those complex phenomena which the social sciences, taken as a whole, seek to explain."

The goal suggested by Mandelbaum is an ancient one. It has been set before scholars in many diverse cultural milieus. Perhaps, in our time, if the boundaries between the intellectual fiefdoms break down and the biological disciplines are structured to interact with all other "sciences of man," we will have a vision of the human condition and human potential to match our knowledge of stars and atoms.

LOGIC, LAWS, AND LIFE

LEONARD J. SAVAGE
1917–1971

The Shifting Foundations of Statistics

> The importance of probability can only be derived from the judgment that it is *rational* to be guided by it in action; and a practical dependence on it can only be justified by a judgment that in action we *ought* to act to take some account of it.
>
> —J. M. Keynes
> *A Treatise on Probability*

1. INTRODUCTION

This essay is a brief survey of the philosophy of statistics by one who is much more a statistician than a philosopher. Statistics is a servant of science, which some might think too humble and humdrum to be of philosophical interest, and indeed the statistician has his share of dreary chores. Yet, since statistics is concerned with the advance of knowledge and with how to behave in the face of uncertainty, it has important contacts with ethics and epistemology. How and whether statistics is to be distinguished from the whole of the philosophy of science in principle is not obvious, though of course the daily work of the statistician is different from that of the philosopher. And there are many topics in the philosophy of science, such as the philosophy of theorizing and of discovery, to which statistics has thus far contributed little if anything. But if the philosophy of statistics is only a part of the philosophy of science, it is at least that—and a part intimately entwined with the rest.

I hope to give you a panorama of the principal views on the foundations of statistics, so far as that is possible within an essay that can be read at one sitting. A wide variety of opinions should be fairly presented, but concealing my own opinion, if I could, would not necessarily be fairer to the others than laying it on the table.

2. WHAT IS STATISTICS?

To appreciate the philosophical problems of statistics, you should of course have considerable prephilosophical experience with statistics; but as this essay is mainly for readers with no such experience, I shall try, as a necessarily weak substitute, to compress into this section some picture of what statistics is.

"Statistics" used to mean numerical information about the state, such as populations, prices, and records of marriages, births, and deaths. Even today many statisticians are particularly occupied with such data. But there is now a sense of "statistics" that has no particular reference to social data but refers to data in general.

To put it a little flippantly, statistics consists in trying to understand data and to obtain more understandable data. This covers much of what statisticians actually do and conveys some truth that more philosophical definitions to be given later will miss. The statistician calls his efforts to understand "the analysis of data" and his efforts to get more understandable data "the design of experiments." People too often ask him for help in analyzing data from studies so ill designed that good analysis is impossible, and too seldom do they invite him to participate in the design.

In principle, statistics is applicable in every area in which there is systematic empirical study. And today the variety of applications in scholarship, business, and government is amazing, as a few examples may illustrate.

An early plan for the famous experiment on the Salk vaccine against polio called for administering the vaccine to second-grade children volunteered by their parents, and comparing the incidence and severity of polio in those treated children with children from the first and third grades. This seemed better than the notoriously bad scheme of comparing volunteered with nonvolunteered children from the same classes; and because the second-grade children tend to be intermediate in age between those of the first and third grades, this design might avoid confounding the effect of age with the putative effect of the treatment. Those ideas are themselves illustrative of statistical thinking, but it came to be recognized that they were altogether too fragile to support a tremendous experiment that was to involve more than a million children and could never be repeated. Therefore, a much more solid design was adopted (at least for a large part of the experiment), in which half of the volunteers received the Salk vaccine and the other half received an innocuous placebo material. In such an experiment, the two groups of children compared are identical except for chance fluctuations, and since neither the child, his parents, the physician administering the vaccine, nor those following the health of the child know whether he is a treated child or a placebo child, there is no possibility of suggestibility being responsible for an apparent effect of the treatment.

Applications of statistics to the humanities are relatively rare, but efforts to determine disputed authorship by counting words and other features of style are a widely known example. For another, I was once consulted on unusual data bearing on the duration of early and modern performances of musical works.

Statistics has, from its beginnings, served the social sciences. The following example from linguistics is interesting. Something can be inferred about the genealogy of related languages and even about their temporal history from the frequency with which the words that represent the same idea in two languages are linguistically cognate, and statistics of course plays a role in any inference based on frequencies.

In dollar-and-cent value and in the total number of people employed, the field of quality control might claim to be the largest part of current statistical practice. "Quality control" means primarily the keeping of a systematic, and often mechanical, watch on manufacturing processes with a view to detecting signs of avoidable trouble.

Some philosophers might expect that the simple and quick way to say what statistics is would be to formulate an accurate, and therefore presumably philosophical, definition. Others would anticipate with me that that approach must at least be balanced with others, if only because a reader would be unlikely to comprehend a succinct and accurate definition—should one be available—without amplifying comment. Still more, actual philosophical definitions tend to be distilled out of controversy and therefore to conceal questions by answering them implicitly. But aphoristic definitions, not necessarily in harmony with each other, may add something to understanding. For example, statistics has been defined as making wise, though uncertain, inferences. Such expressions have a certain value and are not to be scoffed at, but they are useless to one who asks, "What do statisticians do?" because they do not ditinguish between the activities of the statistician and those of a detective or a diagnostician. For many statisticians, something like the following idea, in the words of M. S. Bartlett, is indispensable: "By statistical data and statistical phenomena I refer to the numerical and quantitative facts about groups or classes of individuals or events, rather than facts about the individuals themselves."[1]

3. WHAT IS A GOOD SYSTEM OF FOUNDATIONS OF STATISTICS?

Whether one ought to or not, I make little if any distinction between the philosophy of statistics and the search for a good system of foundations of statistics. Let me say what I think a system should achieve, without pretending that my criteria are either incontrovertible or complete. It should say by implication what constitutes good statistical practice and why, that is, it should give statisticians convincing reasons for doing what they feel sure they should do. This is a little hard to put carefully, for I do not mean that a good theory must comply with all widely accepted practice. In particular, if a theory were to lead to the conclusion that some common statistical practice was wrong, that practice should be carefully reconsidered, and perhaps abandoned, in the light of the theory. But often a theory of statistics leads to some such conclusion that, on reflection, must itself be rejected.

For example, random samples, that is, samples drawn strictly by lot according to some set of rules, are an invaluable tool of modern statistical practice. But the personalistic Bayesian theory of statistics—which I favor—taken literally, denies the utility of randomization. This raises the question whether the theory leads to an untenable conclusion or whether randomization really is contraindicated. The truth seems to be that randomization has sometimes been overrated but is nonetheless a valuable device, and that this theory must be regarded as imperfect because it rejects randomization.

Sometimes adherents of a statistical theory do not seem to regard an absurd prediction as an imperfection; they take the position that, since the conclusion is absurd to a competent statistician, its occurrence is no indication that the theory will lead people astray. A person brought up in the tradition of the crucial experiment will not be satisfied to call an absurd prediction merely an imperfection; he will consider it a disaster that simply disqualifies the theory. For myself, these seem two workable ways of talking about the same state of affairs. Since I despair of ever seeing a theory that says much and yet leads to no wrong conclusions, I am pressed to make use of the practical notion that some theories are better than others and that some may be nearly correct, impossible though it may be to characterize such relative notions in the framework of formal logic.

Apart from having incorrect or absurd implications, a theory of statistics can be regarded as inadequate if there are important aspects of the statistical craft unaccounted for in it.

For example, no systematic theory of statistics known to me seems to have a place in it for descriptive statistics, which is the art of putting data before the human mind in a way that is humanly comprehensible, suggestive of useful ideas, and not misleading, as we do with well-designed charts, graphs, maps, tables, and animated cartoons. Much is known about this art, and much remains to be learned. It involves not only the usual subject matter of statistical theory but also many aspects of human psychology, including the psychology of perception. It is, therefore, empirical in a sense that may exclude it from being the subject of a philosophical or foundational theory, so perhaps a theory of statistics that has nothing to say about descriptive statistics ought not on that account to be scored as inadequate. We must recognize, however, that teaching such a theory and its implications is an inadequate statistical education, because the successful creator or user of statistics needs to know nontrivial things about descriptive statistics. Unfortunately, descriptive statistics has until very recently been in grave disrepute in the classroom, largely because of excessive respect for philosophical and mathematical theories of statistics.

The same trend that has revived descriptive statistics and led to new respect for puttering about with the data in a relatively informal and unstructured way has brought with it an excessive disparagement of formal theory in statistics. It takes experience and judgment to arrive at the right level of formality in a given context. Great clarity and precision without formality seem to be difficult if not impossible, and there are some things that can and should be said with clarity and precision, some that cannot, and others that should not.

As is very well known, every formal theory is completely impotent without an informal interpretation, which can never be nearly so clear and precise as the theory being interpreted. Everyone familiar with the philosophy of science knows that to put the crystalline geometry of the mathematician to everyday use, you have to get down to brass tacks. A simpler and even more striking example is that of mathematical probability. This is so pure and simple a mathematical structure that it serves adherents of widely differing views as to the meaning of what is probable, as each subjects the formal structure to his own interpretation.

There seem to be many things in statistics other than descriptive statistics that must be recognized and talked about but that do not necessarily lend themselves to formalizing. One is the notion, already mentioned, of a nearly correct theory. A very conspicuous one in personalistic Bayesian statistics is that the preferences of a real person are subject to vagueness. Thus, it is an idealization to suppose that you can rank half a dozen proposed menus for tomorrow's dinner in order of preference. Actually, you would probably vacillate in some preferences. And when the choice is not between dinners but between houses, jobs, or spouses, the vacillation can be agonizing. Inevitable and important though this vagueness of preference is, not everyone is agreed that a normative theory of the *homo economicus* can be improved by incorporating vagueness, and formal attempts in this direction have not yet accomplished much, in my opinion.

Whatever we may look for in a theory of statistics, it seems prudent to take the position that the quest is for a better one, not for the perfect one.

4. THE IMPORTANCE OF PROBABILITY FOR STATISTICS

Statistics is about uncertainty. At least, I so understand it for the main purposes of this essay, and it is so understood by most statistical theorists. Perhaps a statistics of the relatively certain is developing. People are increasingly aware of what are called data-rich situations, in which the challenging problem is model-building rather than error analysis. They ask, "What, if anything, does all this bewildering detail mean?" rather than, "Just how sure is the conclusion?" Perhaps this area, which some call data analysis, will someday have a theory and a system of foundations. Indeed, the personalistic Bayesian theory already has some implications for it. But the subject is highly empirical because it is intimately related to descriptive statistics and for other reasons. To be good at data analysis implies among other things having serendipity, and a prescription for that is hardly to be expected.

Returning from a digression, I repeat that statistics is largely, and perhaps in the last analysis exclusively, about uncertainty, and the analysis of probability has always seemed to be a tool—even the main tool—for handling uncertainty.

5. MEANINGS OF PROBABILITY

The differences in theories of statistics are closely associated with differences in attempts to define probability, so I shall briefly discuss the various meanings, or putative meanings, of probability. The ultimate problem here is not really lexicographical. It is not to discover how "probability" is used in certain contexts, nor is it in any way to find out how that word, or any other, ought to be used. To build a system of foundations for statistics, we need to lay hold of any concepts that fruitfully organize our thoughts about uncertainty, and we may anticipate that one or more notions of probability will be among them. For example, utility is a concept very different from any sense of probability, and yet it is almost as essential to thinking about behavior in the face of uncertainty as probability itself. But for everyday statistics, utility is somewhat less important, and I must refrain from discussing it explicitly in this essay.

There have been a great many attempts to define probability, and new attempts and refurbishings of old ones continue. Complicated though the situation is, the myriad attempts to define probability are spun out of relatively few threads, such as the concepts of symmetry, opinion, right thinking, frequency, the almost certain, and perhaps a few others. The riddle is in part to know which of the attempted definitions make sense and which are seriously circular or otherwise intrinsically defective. If a definition does make sense, for what purposes is it useful? Are there other meaningful definitions useful for the same or for other purposes?

Of course, no taxonomy of all the definitions is at once simple and complete, but three main types of definition are generally recognized, which I shall here call necessarian, frequentistic, and personalistic.

I call certain notions of probability necessarian in the sense of logical necessity, because they regard probability as a sort of partial implication. Though in their modern form necessarian theories are based on symbolic logic, I am not alone in regarding them as closely related to the earliest definitions of probability, which arose in the seventeenth and eighteenth centuries when mathematicians first became intrigued by games of chance. Such games invite description in terms of equally likely cases, as when we envisage thirty-six equally likely ways to roll a pair of dice, one red and one green. This notion of equally likely cases seems to arise from the symmetry with which the gambling apparatus is perceived. Sometimes it was adopted with complete naiveté and without any distress over possible difficulty in getting to the bottom of "equally likely." Sometimes a critical effort was made, often by recourse to the idea of "insufficient reason": Any argument leading to the conclusion that three with the red die and four with the green one is more likely than five with the red one and one with the green one could be converted by a mere linguistic change, not affecting the validity of any step, into the opposite conclusion. Whether and when such applications of insufficient reason have validity is a slippery question, but something very like them seems to be at the root of every necessarian theory of probability, old or new.

The nineteenth and twentieth centuries have mainly rejected probability based on symmetry. For when statisticians and actuaries wanted to conclude that the probability of a boy baby is slightly but distinctly more than 1/2, or that the probability that a forty-eight-year-old bachelor will not survive the year is 1.4 percent, they could perceive no "equally likely cases" to serve as a basis for these natural and useful conclusions. Nonetheless, in the twentieth century there has been renewed interest in the possibility of establishing a necessarian concept of probability, led most prominently by John Maynard Keynes, Harold Jeffreys, and Rudolf Carnap.

Modern necessarians would consider it a naive error to look directly for equally likely cases in terms of which to compute probabilities. Rather, viewing the theory of probability as an extension of logic, they seek to define an extent to which a body of evidence logically partially entails a given proposition. On the whole, modern necessarians seem to acknowledge that to do this implies doing it for the special case in which the body of evidence is tautologous. But thus formulating correct opinions for the intelligent but blank mind seems to some of us an implausible task. Necessarians are usually temperate in their claims, acknowledging their theories to be seriously incomplete. Nonnecessarians may not find even the ostensible beginnings of necessarian theories to be cogent.

Few modern statisticians, as opposed to philosophers and others interested in probability, look in the necessarian direction. Harold Jeffreys is the notable exception, and Ronald Fisher might be said to have fallen into a necessarian position with his espousal of "fiducial probability," though he would have denied the imputation.

A frequentist seeks to define probability as the frequency of success in a sequence of trials of "similar" events. There is, of course, some sense in this. When we see that about 51 percent of all new babies are boys—year after year and throughout the world—we are inclined to say that the probability that a new baby will be a boy is about 51 percent. This notion has held the ascendant with statisticians and actuaries, and perhaps with physicists too, during the later nineteenth century and thus far in the twentieth century. The attraction of the frequentistic approach for scientists can be seen in their attitudes toward the two competing approaches. First, the at least seeming hopelessness of earlier symmetry theories has already been mentioned as discouraging necessarian approaches. Second, the personalistic approach was almost voiceless during the period in question, and in particular during the great renaissance of statistics in the early part of the present century. Also the popular, and perhaps plausible, doctrine that science, being concerned with the objective, cannot have its support in subjective judgments has militated strongly against acceptance of the personalistic approach to probability in science.

One great difficulty with ostensibly objective definitions of probability in terms of frequency is the ultimate subjectivity of the judgments of similarity of trials on which application of such definitions must rest. Furthermore, frequency definitions of probability are liable to circularity. To determine empirically whether the probability that this coin falls heads is nearly 1/2, the frequentist typically suggests that we see whether the frequency of heads in a large number of tosses is nearly 1/2. If not, then the probability in question is probably not near 1/2, and otherwise it probably is. But the *probably*s occurring in that criterion are themselves to be based on frequencies, which seems to lead to regression or circularity. Of course, escapes have been suggested, but they do not satisfy everyone.

A personalist views probability as a certain measure of the opinions of an ideally coherent person, and he does not expect each such person to have the same opinions. Personalistic theories are not usually purely mental but rather economic. For example, a vivid, but not fully rigorous, personalistic theory defines your probability of rain this afternoon as the fraction of a dollar you would accept now in lieu of a binding promise to pay you a dollar in case of rain this afternoon. The conditions of coherence usually adduced for such economic behavior entail that the probabilities of the person to whom they apply constitute a system of mathematical probabilities of the usual sort, in technical terms, a nonnegative, additive function on the Boolean algebra of events, that attaches unity to the tautologous event.

A personalistic theory could be given a psychological and empirical interpretation as predicting the behavior of some class of "persons." As empirical theories, they are not very interesting, nor have they very wide domains of validity. Their real importance is as normative theories by which a person, like you, can police himself for coherency. Consider this example.

You are in a classroom with twenty-five strangers. What is the probability for you that at least one pair of your classmates shares the same birthday? Spontaneously, most adults feel that this is rather a longshot, perhaps about one to ten. However, except for a little complication about February 29 and minor adjustments for long winter nights, you may feel vividly that all the 365^{25} possible lists of birthdays are equally probable. By merely counting those among these equiprobable lists in which there is a pair of identical birthdays, it follows that the probability of at least one such pair is more than one-half. Actually, the counting would take more than patience—there have not been 365^{25} microseconds since time was new—and it must be done by a mathematical shortcut. When you find that what at first glance seemed a long shot actually has a high probability in consequence of elementary probabilities about which you feel particularly clear and secure, you, like most others, will perhaps resolve the conflict by deciding that your initial judgment was mistaken and should be amended in conformity with those judgments that continue to seem secure to you. Except for its frivolous content, this example illustrates quite well how the theory of personal probability comes to be of use in practical affairs.

The personalistic view was vaguely outlined in the nineteenth century but mainly only to be repudiated on the grounds that subjectivity had no place in science. It was re-created several times all too quietly between the two world wars by Emil Borel, Frank Ramsey, Bruno de Finetti, and Bernard Koopman. Though these developments were more or less concurrent with the great statistical renaissance of the 1920s, they were a little too late and too ill publicized to have an impact on that movement. Now, however, the personalistic view of probability is the distinctive feature of a school of statisticians, still in the minority, to which I myself adhere.

The personalistic view of probability, like other views, has difficulties and encounters objections. Being a personalist, I am not always confident as to which objections are serious and which are apparent objections only, but I shall try to distinguish them.

Conflict with the vaunted objectivity of science seems an apparent difficulty only, for that objectivity is illusory. We were all brought up on the "known fact" that noble gases could not form genuine compounds with even the most active elements. In recent years, oxides and fluorides of the noble gas xenon have been prepared by the cupful, and we know now that the "fact" was simply an opinion, false but perhaps reasonable. To me, it is evident that all the facts of science are but such opinions. Of course, the opinions that Britain is an island, that the earth is round, and that xenon hexafluoride exists after all are extraordinarily well founded, but I see no hard and fast line separating them from more tentative opinions.

A more serious difficulty with our exploitation of personal probability is this: Perhaps we are too far in our capacities from the ideally coherent man to benefit always by emulating him. Unlike him, we are the victims of vagueness and of relative inability to count and calculate. Indeed, were it not for these human weaknesses, the theory of personal probability would have no human use. Yet, since we are imperfect, we might conceivably be led systematically astray in trying to emulate perfection. An illustration has been mentioned earlier: The perfect *homo economicus,* acting on his own account, would never wish to take a random sample; but random samples are important for us, both when we are acting purely on our own account and when we are acting in the interest of others.

The claim of personal probability to make sense in some contexts is relatively uncontroverted. But is personal probability really useful in statistics and science? The corpus of personalistic Bayesian statistics amounts to a case for the affirmative, but of course it cannot be exhibited within a single essay.

Do other concepts of probability also have a place? Many personalists answer both yes and no: yes, in our respect and understanding for the impulse toward other notions of probability; no, in our conviction that the personalistic notion is an excellent vehicle for expressing what is valid in the other notions, as will now be briefly explained.

Seemingly, necessarian probability arises when people perceive symmetry in their own opinions. If you judge one pair of points to be thrown with a red die and a green one to be as probable as any other, then you necessarily—if you are coherent—judge the probability of a total of seven points to be 1/6. But it seems indefensible to hold that the symmetry of a pair of dice is a fact rather than a judgment; though of course the dice could objectively be very nearly square, very nearly homogenous in weight, and thrown with high velocity into a large square with soft elastic walls, all of which conditions may induce us to judge the thirty-six possibilities to be equally probable. Nor are such judgments of symmetry by any means adequate for all applications of probability.

The role of frequencies has been analyzed thus. A finite sequence of trials is called exchangeable for a person if he regards every pattern of successes and failures to be just as probable as every other pattern with the same number of successes and failures. This might be your own attitude toward successive tosses of a thumbtack that either lands flat on its head or with point and edge in contact with the table. Reflection shows exchangeability to be exactly the kind of similarity that is needed to make frequency of success in past trials all that is relevant to success in future trials. The introduction and study of exchangeable events by de Finetti gave him a very successful way to do what has been less successfully attempted by frequency theories of probability, but this personalistic frequency theory does not pretend to be objective; it is explicitly restricted to situations in which exchangeability describes the user's opinions about a sequence of events. This is akin to situations of perceived symmetry mentioned in the preceding paragraph, because exchangeability is a kind of symmetry.

6. THEORIES OF STATISTICS

The theory of probability that a statistician accepts will have a profound influence on his theory of statistics. The three main views of probability therefore provide a convenient organization for briefly reviewing the prominent theories of statistics.

When all uncertainties are measured by probabilities, as they are for both necessarians and personalists, then statistical inference is in principle very simple. It consists merely in calculating the probability that smoking is at least so dangerous or that this ox yoke is between 2,300 and 2,400 years old on the basis of the available evidence. Such calculations typically make frequent use of an elementary mathematical formula by means of which probabilities of hypotheses in the light of data are calculated, using as ingredients the probabilities of the hypotheses before the particular data in question were seen and the probability of seeing these data under each hypothesis. This formula is called Bayes' theorem, after Thomas Bayes, and it is for the rather shallow reason that both necessarian and personalistic statisticians have relatively frequent occasion to use Bayes' theorem that they are called Bayesian statisticians.

Until recently, Harold Jeffreys was the only prominent necessarian statistician (as opposed to necessarian philosopher) of probability, but now there are a few others, particularly E. T. Jaynes. The objections to their position are basically those to the necessarian approach to probability. We personalists feel that necessarians claim to get something for nothing, and this seems particularly conspicuous in necessarian statistics. Necessarian and personalistic theories of statistics do have important formal similarities, and personalists are deeply indebted to the writings of the necessarian Harold Jeffreys.

The dominant theories of statistics are still frequentistic. They spring from the conviction—I would say, the recognition—that an objective foundation for the probabilities of scientific hypotheses cannot be found and from the conviction—I would say, the illusion—that frequency probabilities have an objective meaning. Frequentistic theories of statistics have usually accepted the conclusion that since initial, or *a priori,* probabilities are ordinarily not available in scientific uncertainties, the result of a statistical inference cannot be a terminal, or *a posteriori,* probability. These theories have tried a variety of devices aimed at accomplishing statistical inference without using the forbidden probabilities. Thus, the difficulties of the frequentistic theories of statistics are not only those arising immediately from the difficulties of defining frequency probability but also those imposed by the rejection of the possibility of describing scientific uncertainty directly in terms of probability.

There are two prominent frequentistic theories of statistics somewhat in conflict with each other, that of R. A. Fisher and that of Jerzy Neyman and Egon Pearson. It is not practical here for me to describe these theories or even to contrast them very intelligibly. One important difference is this: The Neyman-Pearson school, especially as elaborated by Abraham Wald, has increasingly sought escape from its dilemma—that imposed by abstinence from *a priori* probabilities—by replacing inductive inference with economic notions suggested by the phrase "inductive behavior."

Of course, personalistic Bayesian theories of statistics seek to found the theory of statistics on the theory of personal probability. In personalistic Bayesian statistics, as in necessarian Bayesian statistics, the mechanism of inference is relatively simple once the appropriate probabilities are specified. The practical specification of these quantities is perhaps the most characteristic feature of Bayesian statistics. In principle, each probability is an aspect of the opinion of the user; he alone can determine it, by looking inward and in no other way. In practice, it is extremely valuable to note situations in which a wide variety of users will have common opinions, or at least have common features to their opinions. Exchangeability is one of the most important of such features. Others are expressed by willingness to accept members of narrow families of distributions, such as the normal distributions, the Poisson distributions, and the binomial distributions, as appropriate, subject to certain parameters which may themselves have subjective distributions. Incidentally, these special models are never complete descriptions of a serious opinion, but only approximations to it, such as we would express by saying, "I tentatively regard those observations as normally distributed."

Since personal probability, together with utility, provides a framework for decision in the face of uncertainty, personalistic Bayesian statistics can be regarded as an economic, or decision-theoretic, approach to statistics. However, the two concepts of personal probability and utility are theoretical counterparts of opinion and preference, and it is possible, practical, and often useful to study opinion and the change in opinion without explicit reference to values. Thus, Bayesian statistics (of both kinds) is sometimes less self-consciously decision-theoretic than is more recent work in the Neyman-Pearson-Wald tradition.

The person with whom personalistic Bayesian statistics is concerned is not ordinarily the statistician, who is likely to be a consultant serving the person of concern. Or, if the one who consults the statistician is a scientist, he may be largely concerned with putting forth his experience in a form that will help colleagues revise their opinions conveniently and reliably in the light of what he has seen, and it is the opinions of these ultimate consumers that are pertinent.

The relation of personalistic Bayesian theory to other theories is interesting and, for me, one of its most attractive and reassuring aspects. The personalist sees the necessarian as one who hopes for too much; the necessarian sees the personalist as one who hopes for too little. But there is little in their philosophies that would lead them to practice statistics radically differently, except that a necessarian, fascinated by the symmetry or other analytical properties of some probability distribution, may take it seriously in a context in which it could not at all satisfactorily represent the opinions of the person concerned, as does occasionally happen.

Personalistic Bayesian statistics can be viewed as one more step in the development that began with Neyman and Pearson and was carried forward by Wald under the name of statistical decision theory. Yet, the newer theory sometimes gives the appearance of being incompatible with the older one, and some adherents of the older theory strenuously reject the newer. This harmony that sounds like discord is worth expounding upon.

The Neyman-Pearson tradition began with, and long sustained, the credo that statistics must proceed without any recourse to *"a priori"* probabilities and, more generally but almost consequently, without attaching probabilities to individual events. Adherents of such a theory are therefore inclined to be averse to one that makes free use of the forbidden probabilities. But—and this seems true, however surprising—the frequentistic decision theory when fully developed leads inexorably back to the originally renounced probabilities. This surprising development cannot be properly explained here, but can be hinted at.

The Neyman-Pearson theory has always stressed that many choices in its application are ultimately to be made by the user, so these theories have always had room in them for the expression of subjective opinions. Yet, in view of the objectivistic attitude of this theory toward probability, it might be anticipated that the freedom it allows the individual user would be less than that allowed by personalistic Bayesian theories—in fact, there has been some temptation to call the more traditional theories objectivistic and the personalistic Bayesian theory subjectivistic—but in reality the personalistic Bayesian theory does not allow the individual as much freedom of choice as the older one does and subjects him to considerable additional discipline. Adherents of the older theory do sometimes object to the new theory as introducing subjectivism in science, which in their view should be purely objective, but they sometimes vehemently complain that the new theory regiments what they conceive to be legitimate personal freedom. We personalists are, however, glad to give up the freedom to be incoherent, especially when we reflect on its economic interpretation.

Another obstacle to acceptance of the newer theory is that it does contradict many prominent Neyman-Pearson traditions, but these, according to us, are already in conflict with mature Neyman-Pearson theory, especially that since Wald. Prominent examples of notions rejected by personalistic Bayesian statistics are confidence intervals, tail-area tests, and strictures against optional stopping.

The Fisherian tradition, though far less widespread than that of Neyman and Pearson, is even more elusive, so it is hard to say to what extent personalistic Bayesianism blends, and to what extent it clashes, with the Fisherian tradition. A beautiful, important, and, I think, reliable principle of statistics called the likelihood principle was first put forward by Fisherians and can be arrived at by diverse routes; it is an immediate corollary of any Bayesian theory but is in striking contrast with some forms of the Neyman-Pearson tradition. On the other hand, the personalistic Bayesian position seems to have no room for the notion of fiducial probability to which Fisher attached considerable importance. Nor has it any interpretation for the tail-area tests of sharp null hypotheses, which Fisherians interpret somewhat differently from Neyman-Pearsonians, but which both emphasize.

7. DO THE FOUNDATIONS MATTER?

When I first became interested in the foundations of statistics, my natural bent was to expect no impact at all on statistical practice from study of the foundations. Will large numbers of intelligent people intent on their business and in good contact with each other persist in mistakes that could be detected by thought alone? My Panglossian personality, holding that all must be well in this best of all statistical worlds, was simply curious to understand why. And, indeed, many errors to which traditional statistical theories led were repeatedly corrected in the practice and publications of good statisticians, but the newer theory seems to conflict much less with practice and, therefore, to provide a much better vehicle for self-discipline, for instruction, and for communication with those whom statisticians serve.

8. THE FUTURE

The foundations of statistics are shifting, not only in the sense that they have always been, and will doubtless long continue to be, changing, but also in the idiomatic sense that no known system is quite solid.

The personalistic Bayesian position is gaining in adherents and having an influence even among those who do not adhere to it. This may well be the wave of the immediate future in statistics. If so, it is likely to be dangerously oversimplified and lead to license in its application, which will provoke some sort of restraining reaction.

About the less immediate future, who knows? Many important topics, to some of which I have already referred, do not fit well into any systematic theory of statistics known to me, for example, the cost of calculation as one of the costs to be reckoned with in scientific work; vagueness; the logic of discovery and serendipity; and descriptive statistics. These topics merit, and will continue to receive, philosophical discussion, and they may give the impetus to substantial new theories of statistics.

NOTE

1. Leonard J. Savage et al., *The Foundations of Statistical Inference: A Symposium* (New York: John Wiley and Sons, 1962), p. 38.

REFERENCES

The annotated references below lead to further reading in various directions suggested by this essay. Some of these references have extensive bibliographies of their own. The one in my *Foundations of Statistics* is briefly annotated.

Birnbaum, Allan. "On the Foundations of Statistical Inference." *Journal of the American Statistical Association,* 57 (1962): 269–306.

An interesting derivation of the likelihood principle from a particularly neutral standpoint.

Edwards, Ward; Lindman, Harold; and Savage, Leonard J. "Bayesian Statistical Inference for Psychological Research." *Psychological Review,* 70 (1963): 193–242.

A fairly complete but relatively amathematical introduction to practical personalistic Bayesian statistics.

Ellsberg, Daniel; Fellner, William; and Raiffa, Howard. "Symposium: Decisions Under Uncertainty." *Quarterly Journal of Economics,* 75 (1961): 643–94.

Discusses an important line of dissent within the personalistic position.

de Finetti, Bruno. "Probability: Interpretations." In *International Encyclopedia of the Social Sciences,* vol. 12, pp. 496–504. New York: Macmillan Co. and The Free Press, 1968.

A brief but deep discussion of the relations among the various attempts to define probability.

Hacking, Ian. *Logic of Statistical Inference.* Cambridge: Cambridge University Press, 1965.

Unsurpassed for its broad treatment of the philosophy of statistics.

Hildreth, Clifford. "Bayesian Statisticians and Remote Clients." *Econometrica,* 31 (1963): 422–39.

Clarifies the centrally important situation in which the Bayesian person is John Q. Public.

Jaynes, Edwin T. "Prior Probabilities." *Institute of Electrical and Electronic Engineers, Transactions on Systems Science and Cybernetics,* vol. SSC-4, no. 3 (1968): 227–41.

A good introduction to the author's unusual necessarian position.

Jeffreys, Harold. *Theory of Probability.* 3rd ed. Oxford: The Clarendon Press, 1961.

A recent edition of a masterpiece precious to all Bayesians, necessarian and personalistic.

Kyburg, Henry E., Jr., and Smokler, Howard E. *Studies in Subjective Probability.* New York: John Wiley and Sons, 1964.

An invaluable anthology representing Borel, Ramsey, de Finetti, Koopman, and others, with extensive bibliography.

Lehmann, Erich L. "Significance Level and Power." *Annals of Mathematical Statistics,* 29 (1958): 1167–76.

Pinpoints where the Neyman-Pearson theory stops short.

Salmon, Wesley C. *The Foundations of Scientific Inference.* Pittsburgh: University of Pittsburgh Press, 1966.

Reviews the inference problem from Hume to the present day with considerable reference to the philosophy of statistics.

Savage, Leonard J. *The Foundations of Statistics.* New York: John Wiley and Sons, 1954.

Consists in part of an axiomatic study of personal probability and utility, merging ideas of de Finetti and of von Neumann and Morgenstern. The author, though interested in personal probability, was not yet a personalistic Bayesian and was unaware of the likelihood principle. Has a useful bibliography.

———, et al. *The Foundations of Statistical Inference: A Symposium.* New York: John Wiley and Sons, 1962.

A lecture by me introducing lectures and discussion by others. Very valuable because of the give-and-take.

Wolfowitz, Jacob. "Bayesian Inference and Axioms of Consistent Decision." *Econometrica,* 30 (1962): 470–79.

Dissent from the thesis that personalistic Bayesian statistics is a natural completion of frequentistic decision theory.

RONALD N. GIERE
Indiana University

Testing Versus Information Models of Statistical Inference

> The scientific man finds himself confronted by phenomena which he seeks to generalize or to explain. His first attempts to do this, though they will be suggested by the phenomena, can yet, after all, be reckoned but mere conjectures; albeit, unless there be something like inspiration in them, he never could make a successful step. Of those conjectures—to make a long matter short—he selects one to be tested. In this choice, he ought to be governed solely by considerations of economy. If, for example, the prospect is that a good many hypotheses to account for any one set of facts, will probably have to be taken up and rejected, and if it so happens that, among these hypotheses, one that is unlikely to be true can probably be disposed of by a single easy experiment, it may be excellent economy to begin by taking up that. In this part of his work, the scientist can learn something from the business man's wisdom.
>
> —Charles S. Peirce
> *"Lessons from History of Scientific Thought," 1901*

1. INTRODUCTION

Until about 1960, the major influences on the course of contemporary Anglo-American philosophy of science came from developments in physics, that is, relativity theory and quantum mechanics, and in mathematics, especially mathematical logic. Both reached a definitive, though of course not final, form during roughly the first third of this century. These developments have had a profound effect not just on the philosophy of physics and mathematics, but on all areas of the philosophy of science, including the subject commonly labeled "probability and induction." Until very recently, most philosophical work on the nature and justification of scientific claims to knowledge strongly reflected the general influences from physics and mathematical logic. On the one hand are discussions of the hypothetico-deductive method as applied to the grand theories of physics from Newton's mechanics to Einstein's. The dialectic of conjecture and refutation propounded by Karl Popper and his followers is in this tradition.[1] On the other hand is a direct line of development from *Principia Mathematica* through John Maynard Keynes and Rudolf Carnap to Jaakko Hintikka and his school. Much of this work, under the title "confirmation theory," does not even treat the simplest universal generalizations, let alone grand physical theories.[2] Of course both traditions include the claim that their approach already encompasses, or will eventually encompass, everything worthwhile in the other tradition. In neither case, however, is this claim backed up with much beyond faith in one's own program.

The first third of the century witnessed other developments in science and mathematics that could have been taken as the starting point for developing a richer and more comprehensive scientific epistemology, namely, genetics and mathematical statistics. It is interesting to speculate on the personal, intellectual, and social reasons why these developments did not have a more substantial impact on the philosophy of science. But this is not my purpose here. My purpose is, roughly, to do now what was not done then. This way of stating the task, however, is too simple. It suggests that one might now just generalize a bit on current applications of mathematical statistics to biology and thereby achieve a comprehensive account of scientific inference. In fact there was a brief period just before and just after World War II when this seemed possible and, indeed, was attempted by R. B. Braithwaite and by C. West Churchman.[3] But the situation is now more complex.

Since the mid 1950s it has gradually become clearer that there are two competing paradigms of statistical inference, and of scientific inference generally. The expressions "testing paradigm" and "information paradigm" are as accurate as any. By a "paradigm" of scientific inference I mean a general conception of the kind of process scientific inference is or should be. Specific theories of inference may be thought of as "models" or "interpretations" (not in the technical sense obviously) of the more general background paradigm. The role of the paradigm in the development of specific models of inference is that of inspiration and guide. There is no such thing as the uniquely right interpretation of a paradigm. For example, R. A. Fisher's significance tests and Jerzy Neyman's confidence intervals, as well as traditional hypothetico-deductive methods, may all be viewed as interpretations of the testing paradigm. Carnap's 1950 system of logical probability, L. J. Savage's 1954 theory of personal probability, and the recent relative likelihood models of Ian Hacking and A. W. F. Edwards may be seen as interpretations of the information paradigm.[4] The paradigms influence, but do not determine, a choice among interpretations.

It is difficult to say very much about the paradigms themselves. They are inherently vague, exceedingly general, and largely invisible, even to those who hold them. The information-processing paradigm is probably the easier of the two to describe. It is also the one that contemporary philosophers, especially logicians, find most intuitively satisfying. The leading idea of the information paradigm is that there should be a direct measure of the bearing of evidence on hypotheses. As new information comes in, whether in the form of experimental data or not, it is to be processed more or less continuously. The output of the information-processing is, of course, a new valuation of all relevant hypotheses. In the classical subjective models of F. P. Ramsey, Bruno de Finetti, and Savage, the output is a rational degree of subjective belief in the hypotheses, and information is utilized through a process of conditionalization, that is, Bayes' theorem.[5]

In the testing paradigm there is no direct evaluation of hypotheses relative to data. Rather, the data are the output of a setup designed to test a particular hypothesis. The result of the testing process is not a reevaluation of some continuous measure, but a simple dichotomous acceptance or rejection of the hypothesis under investigation. A hypothesis is not accepted because it has a high value (on some scale) relative to the information. It is accepted because it has passed an "appropriate" test, for example, one with an appropriately low probability for rejecting a true hypothesis and perhaps also an appropriately high probability for rejecting a false one. In general, the testing paradigm makes scientific inference a series of discrete and relatively definitive, though fallible, processes—not a continuous process of reevaluation relative to an incoming flow of information.

Both the statistical and philosophical literatures contain numerous sweeping dichotomies, for example, subjective versus objective, empirical versus logical, testing versus estimation, inference versus decision. My distinction between testing and information-processing paradigms cuts across these dichotomies. Perhaps I have misperceived the basic paradigms. Maybe there are more than two. This is possible, but it is fruitless to debate at such a high level of generality. To grasp the paradigms one needs to look first at some particular models of statistical inference.

In part 2 of this essay I will outline the conceptual development of testing models of statistical inference from roughly 1900 to 1950. This sketch will fail to be historical, not, I hope, through inaccuracy, but simply through the omission of so much that is historically important. In part 3 I will take up an example of a common type of objection to modern testing models of statistical inference. I will argue that this type of example does not provide an objection to testing models as such. Rather it only calls into question the identification of empirical probabilities with limiting relative frequencies, an identification that has invariably been associated with testing models of inference. The objection can be overcome, I argue, by adopting even a mild form of propensity interpretation of probability. Part 4 will be devoted to an exposition and criticisms of some arguments of Dennis Lindley and L. J. Savage to the effect that modern testing models are either incomplete or inconsistent and can be completed only by converting them into Bayesian information models. Although these are among the most serious objections raised against contemporary testing models, they have received very little critical attention. I will argue that Lindley and Savage have proved much less than they intended, but that they have given a particularly forceful view of the kind of incompleteness that has characterized testing models for a generation. In part 5 I sketch a way of completing some standard testing models in two specific types of scientific investigation, namely, when testing simple causal hypotheses and when testing complex causal theories. The needed ingredients, I will argue, can be provided by the objective methodological requirements of these scientific contexts together with a restricted decision principle commonly called "satisficing." The conclusion, part 6, consists of some brief remarks on the kind of connection between scientific inference and practical decision-making implied by my analysis of statistical inference in scientific contexts.

2. TESTING MODELS OF STATISTICAL INFERENCE: FISHER, NEYMAN-PEARSON, AND WALD

One can find recognizable tests of statistical hypotheses in the work of Laplace and even earlier. Nevertheless, the mainstream of the tradition that produced contemporary testing models of statistical inference began with Francis Galton, was developed by Karl Pearson and his associates, and was first consolidated by R. A. Fisher. The interactions of both men and ideas in the areas of biology and statistics during the half century between 1875 and 1925 would surely make a fascinating history—one that is just beginning to be written. Galton was a cousin of Darwin's, and a pioneer in studies of human inheritance. He discovered a now common correlation coefficient and was the principal financial backer of *Biometrika,* Pearson's journal. Pearson further developed Galton's correlation coefficient, discovered the chi-squared test of goodness of fit, was a leading figure in the eugenics movement, and wrote the definitive biography of Galton. Finally, Fisher, who more than anyone else founded the discipline of mathematical statistics, began as Pearson's associate, though they later became and remained bitter antagonists.

And it was Fisher who, along with J. B. S. Haldane and Sewell Wright, reconciled natural selection with Mendelian genetics. However, here I am more concerned with the logical development of models of statistical hypothesis testing than with the historical interplay of men and ideas. For this purpose it is sufficient to begin with Fisher's account of tests of significance.

2.1. Fisherian Significance Tests

Fisher never did publish a systematic account of the theory of significance tests. What one finds, rather, are expository comments and numerous examples. The most accessible sources of both comments and examples are the three volumes *Statistical Methods For Research Workers* (1925), *The Design of Experiments* (1935), and *Statistical Inference and Scientific Method* (1957). The latter volume contains the most systematic theoretical discussion of significance tests, but it was written after the main impact of Fisher's work had already occurred. The earlier volumes went through many editions (fourteen for the earliest one), and the most recent contain revisions that clearly reflect Fisher's later thoughts on statistical inference.[6]

It is appropriate to begin, as Fisher usually does, with an example. Consider the classical Mendelian theory for one gene, simple dominance, equal viability, and so forth. Assume a parental generation in which each mating consists of one homozygotic parent from each pure strain. The first filial generation, F_1, will consist solely of heterozygotes, each exhibiting the dominant phenotypic trait but carrying a recessive allele. The following, F_2, generation should then exhibit the familiar 3 : 1 ratio of dominants to recessives. This process has traditionally been represented as in figure 1. Here A represents the allele, Mendel's factor, associated with the dominant trait, and a the factor associated with the recessive trait. Simple dominance means that only offspring with two recessive factors exhibit the recessive trait.[7]

Now suppose we wish to determine whether some particular trait, for example, eye color in fruit flies or skin texture in sweet peas, is transmitted according to the classical Mendelian model. The question then arises whether this trait occurs with probabilities 3/4:1/4 in the F_2 generation.

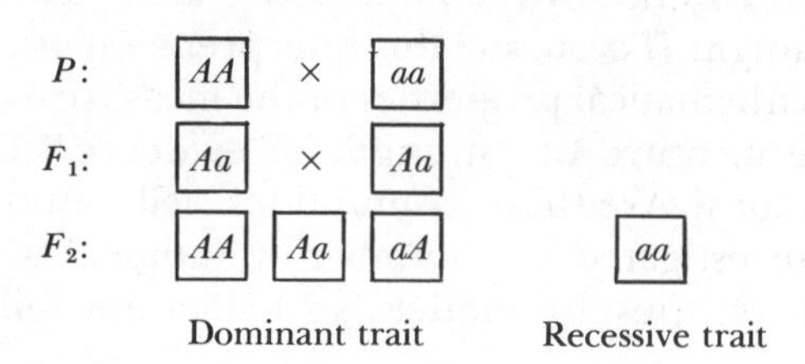

Figure 1

But of course any breeding experiment is unlikely to produce F_2's in exactly a 3 : 1 ratio. Thus, the question becomes one of how close or how far from the Mendelian ratios the observed ratio must be to count for or against the hypothesis of simple Mendelian transmission. Answering this question is the job of a test of significance.

Suppose we attempt to test the hypothesis of simple Mendelian transmission (H) by observing a (relatively small) sample of 20 F_2's. Assuming the order of births is irrelevant, there are 21 possible outcomes of the experiment corresponding to the possible numbers of sample F_2's exhibiting the dominant trait. Now it is a simple matter to calculate the probability of each possible outcome assuming simple Mendelian transmission. From these calculations we can see that the probability of obtaining 11 or fewer dominants is just a bit over 4 percent if H is true. Thus, if the experiment were to result in 10 dominants, Fisher would regard this as being significantly different from the expected value of 15 dominants. On this basis he would recommend that we provisionally reject the hypothesis of simple Mendelian transmission for the trait in question.

The problem for Fisher and later students of statistical inference has been to give a more detailed and comprehensive account of this inference from the low probability of the observed result relative to H, to the provisional conclusion that H is false. Fisher's own writings contain at least two suggestions for a more well-developed interpretation, neither (unfortunately) in sufficient detail. The interpretation that seems best to represent his most mature thought is more in keeping with an information paradigm than with a testing paradigm. Thus, in *Statistical Methods for Research Workers* he writes: "The actual value of P obtainable from the table by interpolation indicates the strength of the evidence against the hypothesis. A value of Chi-squared exceeding the 5 per cent point is seldom to be disregarded."[8] The probability distribution referred to is chi-squared rather than a simple binomial, but the logic of the inference is the same. Moreover, the reference to 5 percent as the boundary for statistically significant results is explicit. The crucial idea is that the low probability of the observed outcome relative to the hypothesis is regarded as a *measure* of "the strength of the evidence against the hypothesis." The existence of such a measure is the leading idea of the information paradigm. To pursue this interpretation one would have to investigate the mathematical properties of the measure to see whether or not it is a viable measure of "strength of evidence." Does it require unstated assumptions? Are these assumptions well founded? Such questions have been investigated by a number of authors.[9] My primary concern, however, is with testing models, so I shall not follow this line of development.

It is surprisingly difficult to find clear testing interpretations of significance tests in Fisher's writings. The following passage from *The Design of Experiments* is among the best I know:

It is usual and convenient for experimenters to take 5 per cent as a standard level of significance, in the sense that they are prepared to ignore all results which fail to reach this standard, and, by this means, to eliminate from further discussion the greater part of the fluctuations which chance causes to have introduced into their experimental results.[10]

The following parallel passage comes not from Fisher, but from a close collaborator and follower, Kenneth Mather. Presenting the idea of a significance test in the introduction to *The Measurement of Linkage in Heredity,* he writes:

The level of probability chosen as indicating significant departure from hypothesis is simply the level at which the worker is willing to be misled. If, as is usual, the one in twenty level is taken, he will find that his supposedly real departures are actually chance ones, once in twenty cases. . . . If this is constantly borne in mind the experimenter will set his levels of significance to suit his circumstances and will not be disconcerted when an apparently promising line of work comes to nothing because it was based on a false conclusion as a result of his test of significance misleading him.[11]

No talk here of the strength of evidence. Rather the focus is on a procedure for sorting experiments into the class of those one is "prepared to ignore" and those one is not. The significance level is the maximum probability that the sorting procedure has "misled" one into rejecting as false a null hypothesis that was in fact true. In addition, there is the interesting suggestion in Mather's paragraph that the level one sets, for example, 5 percent or 1 percent, depends roughly on how willing one is to see "an apparently promising line of work [come] to nothing." Here one gets a glimpse of the kind of utilities that might be involved in a decision-theoretic account of significance tests. But that comes later.

There are many problems with this sketchy testing interpretation of significance tests. Two, however, are especially important because of their connections with later developments. The first is: What do you do if you fail to find a statistically significant difference? In particular, can you accept the test hypothesis as true? Fisher was adamant in opposition to any such interpretation. "Every experiment," he wrote, "may be said to exist only in order to give the facts a chance of disproving the null hypothesis."[12] If the test hypothesis is not rejected, nothing further can be concluded. Now one might think that if *H* is rejected as false, surely its negation, not-*H,* must be accepted as true. So some hypotheses may be accepted. But Fisher denies even this on the ground that not-*H* is not an "exact" hypothesis and therefore could not be the object of a significance test because it could not by itself determine a unique probability distribution over the possible outcomes of any experiment. This is all true, but not really relevant, because by Fisher's own words, no test hypothesis is ever accepted anyway. On the other hand, there is a certain coherence to Fisher's position. If a significance test could lead one to accept the test hypothesis as true, there ought to be some thought given to the probability that one has been misled into accepting a false hypothesis. As it is, a significance test only takes into account the probability that one has been misled into rejecting a true hypothesis. So within the confines of the logic of significance tests, Fisher's position makes some sense. But it is also possible that these confines are too narrow.

The second and more fundamental problem is that we are given no systematic criterion for determining precisely which set of outcomes with low probability is to be the critical set leading to the rejection of *H*. Looking again at the Mendelian example, why not remove the outcome 11 dominants from the critical set and include the outcomes 19 and 20 dominants? The resulting set still has a probability of occurrence less than 5 percent if *H* is true. The same point can be made logically more dramatic by considering that in a sample of 1,000, the probability of getting exactly 750 dominants would be less than 5 percent. Yet surely one would not reject *H* if one did obtain exactly 750 dominants. So there is something wrong with including the most probable outcome in the critical set. But what exactly is wrong with it?

In the Mendelian example there is a clear scientific reason why one might want the outcomes of 19 and 20 dominants in the critical set. The trait might not follow the Mendelian laws, but instead have a very high probability for dominants. If this were the case, there would be almost no chance that the observed outcome would appear in a critical set containing just those outcomes with 11 or fewer dominants. Again, if one were to choose a contiguous critical set containing the most probable outcome, for example, 750 out of 1,000, one would be in the absurd position of being *less* likely to reject *H* the further from the Mendelian value the true value happened to be. In short, in considering various possible critical sets for the rejection of a test hypothesis, one is naturally concerned not only with the probability that one is misled into rejecting a true hypothesis, but also with the probability that one might be misled into not rejecting a false one. Moreover, if at this point one equates nonrejection with acceptance, then the problem of what if anything can be accepted and the problem of choosing a critical rejection set merge nicely into one.

This brief discussion is sufficient to suggest an important conclusion. Fisher's account of significance tests yields at best an incomplete interpretation of the general testing paradigm applied to statistical hypotheses. A better interpretation would include explicit criteria for the choice of a rejection set and provisions for considering the probability of not rejecting a false hypothesis. These are just the elements introduced by Jerzy Neyman and Egon Pearson (Karl Pearson's son!) in the late 1920s and early 1930s.

2.2. Tests of Statistical Hypotheses: The Neyman-Pearson Model

When Neyman and Egon Pearson began to develop the idea of a "best test" in the mid- to late 1920s, they, like Fisher, had not firmly committed themselves to either the information paradigm or the testing paradigm. By the mid 1930s the ambiguity had been resolved, and the result was a well-articulated testing model of statistical inference.[13]

Some of the basic elements of the resolution were points they inherited from Fisher. Most fundamental was Fisher's rejection of the Laplacean model (clearly an information model) which required assigning probabilities to hypotheses. In its place was the doctrine, derived from a tradition including John Stuart Mill, John Venn, and Karl Pearson, that probabilities are to be understood in terms of actual, or at least hypothetical, relative frequencies. The reason Neyman and Pearson took these concepts beyond Fisher is that they were less willing than he to leave the gaps in the model to be filled in by scientific intuition alone. They set out to fill the gaps with conceptual clarity as well as mathematical rigor.

The average contemporary student of classical (non-Bayesian) statistics faces an amorphous collection of techniques derived from Fisher, Neyman, and later writers. In addition to hypothesis testing, these include methods of estimation, regression analysis, analysis of variance, and many other techniques of "data analysis." It is difficult to find much theoretical unity in all this because, in fact, there is little to find. Yet there is a solid theoretical core to the Neyman-Pearson account of statistical hypothesis testing, and this is worth examining because it is the best-developed example of a testing model of scientific inference.

Fisher often failed to distinguish statistical hypotheses from simple causal hypotheses of wider scope. Neyman and Pearson explicitly characterize and distinguish two classes of statistical hypotheses. A *statistical hypothesis,* for Neyman and Pearson, is any hypothesis that assigns a probability distribution to an actual or hypothetical set of events. In the Mendelian example, it is a statistical hypothesis that the probability of an offspring exhibiting the dominant trait is .75. This hypothesis implies a unique probability distribution over the possible numbers of dominants in a sample of 20. Statistical hypotheses yielding unique distributions are called *simple;* all others are *composite.* The statistical hypothesis that the probability of dominants is something other than .75, for example, is clearly composite.

The idea that in designing a test one considers other hypotheses besides the test hypothesis is made explicit and precise by the concept of an *admissible hypothesis.* Just as any experiment has a set of possible outcomes, only one of which will be actual, so there is assumed to be a set of possible hypotheses (the admissible set), one and only one of which is true. The admissible set is partitioned into two nonempty subsets. One set is designated the test hypothesis, the other is the alternative. Formally this designation is arbitrary, though in any real application it clearly is not.

The set of possible outcomes of any experiment, the *sample space,* is likewise partitioned into two nonempty subsets. One is designated the *rejection region,* with the intended meaning that the test hypothesis is to be rejected as false if the outcome of the experiment occurs in this region. Again the partition is formally, though not scientifically, arbitrary.

We are now in a position to give a formal definition of the general structure of a Neyman-Pearson test of a statistical hypothesis (the intended meanings being in parentheses). GNPT = $<A,H,S,R>$ is a generalized Neyman-Pearson test of the statistical hypothesis H iff:

1. A is a set of statistical hypotheses (the admissible set).
2. H (the test hypothesis) is a proper subset of A.
3. S is a sample space (the possible outcomes of the experiment).
4. R (the rejection region) is a proper subset of S.

Such a general definition, however, is useless without criteria for "good" or in some sense "optimal" tests. This requires more interpretation. It is at this point that allegiance to the testing paradigm becomes crucial, for a good test is not defined in terms of an information measure relating the test hypothesis and the data. Rather it is defined in terms of characteristics of the test itself.

Neyman and Pearson begin with a testing interpretation of Fisher's significance level, that is, as the probability that the test leads to the rejection of the test hypothesis when it is in fact true. Such a mistake is dubbed Type I error. In addition they explicitly recognize the possibility of Type II error, that is, not rejecting the test hypothesis when it is in fact false. Thus, in addition to the sample size, n, and the significance level, α, there is a third parameter, β, the probability of Type II error. Any specification of a good test will have to take account of all three.

It is obvious, of course, that a good test should have "low" probabilities for both types of errors. But if several different rejection regions give equally low error probabilities, is there some basis for claiming that one is better or, indeed, best? Clearly the admonition simply to minimize both types of error is not well formulated. Both can be made arbitrarily low if n, the sample size, is allowed to increase indefinitely. And if n is fixed, one needs an additional criterion to determine when both have been optimized. These questions, however, cannot be attacked without assuming a more specific structure for the set of admissible hypotheses. Here the distinction between simple and composite hypotheses provides a natural threefold classification: simple hypothesis versus simple alternative; simple hypothesis versus composite alternative; composite hypothesis versus composite alternative.

Criteria for a best test are most easily justified for tests of a simple hypothesis versus a simple alternative. While such tests are hardly the most common type found in actual scientific practice, they are not merely academic. One might, for example, wish to distinguish two different Mendelian models for the transmission of a particular trait, say a one-gene and a two-gene model. Each may predict a unique but different probability for the trait, for example, 1/2 and 9/16, and there may be good reasons for thinking that these are the only two possibilities. For such a case Neyman and Pearson proved the following simple and powerful result. Let the rejection region be determined by the condition:

$$\text{(LR)} \qquad R = \{x\colon P_K(x)/P_H(x) \geqslant c\},$$

where x is a point in the sample space and c is determined by the condition that $P_H(\mathrm{X} \in R) \leqslant \alpha$. It follows that if R' is any other rejection region with $\alpha' \leqslant \alpha$, then $P_K(\mathrm{X} \in R) > P_K(\mathrm{X} \in R')$. What this means is that if the sample size is fixed, condition LR (for likelihood ratio) yields the rejection region with the lowest possible Type II error for which the probability of Type I error is no greater than α. The probability of *not* making a Type II error, $P_K(\mathrm{X} \in R) = 1 - \beta$, is known as the *power* of the test against the alternative K. Another way of stating the above conclusion, then, is that (LR) determines *the most powerful* α *level test* of H versus K. This result is often referred to as "the fundamental lemma of Neyman and Pearson."

The most common testing situation is one in which the appropriate test pits a simple test hypothesis against a composite alternative. The Mendelian example is like this. The original question is whether some particular trait follows the simplest Mendelian model. This leads to, though it is clearly not the same as, the question whether the probability of dominants in the F_2 generation is .75 or not. The simple test hypothesis is $H\colon p = .75$. The composite alternative is $K\colon p \neq .75$. Since H is simple, the probability of Type I error, $P_H(\mathrm{X} \in R)$, is determined for any region R. But the power, $P_K(\mathrm{X} \in R)$, is not well defined without a weighting function over the simple components $k \in K$. Such a weighting, however, would be equivalent to a prior distribution of probabilities over this set of simple hypotheses. Except for some especially contrived situations, it is hard to give an objective interpretation to these probabilities. What, for example, is the prior probability that F_2's will exhibit the given trait with probability .81? Even if one could give an objective interpretation of what this probability might be, its value is surely unknown. Finally, of course, if such probabilities were available, one could easily develop an information model of inference for these situations. The problem, then, is to fill in the testing model without employing a distribution of prior probabilities over the components of a composite alternative.

Although this is not scientifically very realistic, suppose for the moment that the only admissible hypotheses for the probability of the dominant trait among F_2's were $p \geqslant .75$. Thus, the composite alternative, K', would consist of simple hypotheses corresponding to all values of p such that $.75 < p \leqslant 1$. For any one of these simple components we can use (LR) to construct the best test, that is, the best rejection region, given fixed α. The same region, however, is likewise best against every simple $k \epsilon K'$. Such a test is said to be a *uniformly most powerful* (UMP) test of simple H against composite K'. With the sample size and significance level fixed, such a test is obviously optimal.

Unfortunately only restricted sets of admissible hypotheses yield UMP tests. For K: $p \neq .75$ there is no UMP test. The rejection regions best against alternatives with $p > .75$ are not best against alternatives with $p < .75$. A new criterion for "good" tests is needed. In this case a plausible criterion is that the test be "unbiased," that is, for all $k \epsilon K$, $P_k(X \epsilon R) > P_H(X \epsilon R)$. Intuitively, it should always be more probable that the test rejects H if it is false than if it is true. With this constraint in force, the rejection region is determined by (LR) applied to the remaining possibilities. The power of the test may then be represented as a function of the possible values of p as in figure 2. If one understands this test it should be obvious why the lowest point on the curve corresponds to the significance level, α.

For some composite alternatives to a simple test hypothesis there are neither UMP tests nor UMP unbiased (UMPU) tests. In such cases there are two choices. Either one introduces additional criteria which yield a uniquely best test, or one simply insists that no best test exists for such sets of hypotheses. There is no *a priori* reason why there must be a best test for every possible simple hypothesis and composite alternative.

Similar remarks are in order when both the test hypothesis and the alternative are composite. In this case neither error probability is well

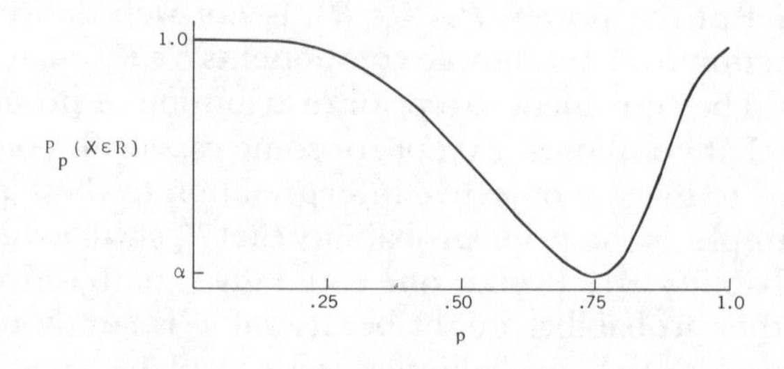

Figure 2

defined in the absence of a prior probability distribution over the admissible set. The most one can give are upper and lower bounds. Neyman has proposed a criterion, analogous to (LR), for use in such cases, but it seems neither widely used nor discussed.[14] Most practitioners try hard to avoid considering such tests.

I have said enough, I hope, to show that the Neyman-Pearson model of statistical hypothesis testing provides a clear example of a model within the general testing paradigm. There is no direct measure, and certainly no probability measure, of the bearing of statistical data on any hypothesis. Rather, the measures, in the form of error probabilities, are measures of the goodness of tests. Hypotheses are judged only insofar as they pass or fail to pass a suitably good test.

It is also clear that there are serious questions as to how complete or comprehensive the Neyman-Pearson model really is. It might be held that a complete model of statistical hypothesis testing would determine a best test of any hypothesis, H, against any alternative K, whether simple or composite. The Neyman-Pearson model is not this comprehensive. There are possible admissible sets for which no intuitively reasonable criteria for a best test have been proposed, and others for which the reasonableness of proposed criteria is in question. Nor is it clear what makes a criterion intuitively acceptable. Thus, there is no definite basis on which one could claim that there are no good tests for certain kinds of admissible sets. All that can be claimed with confidence is that there are some important kinds of admissible sets, for example, simple H versus composite K, for which clearly optimal tests, for example, UMPU tests, do exist.

Historically the main criticism of the Neyman-Pearson model has not been that it is insufficiently comprehensive. It has been that there is confusion over the basic understanding of what the end product of a test really is. This line of criticism, in fact, was begun by Neyman himself.

It is natural to think that if a test leads to the rejection of H, we "infer" or "conclude" that K is true. Yet from the late 1930s on, Neyman has insisted that there is no such thing as inductive inference or inductive reasoning.[15] All "reasoning," he insists, is deductive. There is, however, "inductive behavior." That is, depending on the outcome of a test, one may decide to behave in one way or another. And deciding to behave in a certain way is not an act of reasoning but an act of will. As an example of this viewpoint, suppose in the Mendelian case that the trait in question is the height of a particular kind of plant. And suppose that the value of plants of this type is such as to make it economically worthwhile breeding them if the F_2 ratio of dominants is at least .75, but not if it is lower. Here it would be reasonable to set up a UMP test of $H: p = .75$ versus $K: p < .75$. The possible results of the test, however, are not that one accepts or rejects H, but rather that one begins breeding these particular plants or not. This is a clear and understandable bit of behavior—unlike the "action" of accepting or rejecting a hypothesis.

By assimilating the acceptance or rejection of a hypothesis to choice of one or another course of practical action, Neyman implicitly suggested a way of solving the main problem of *applied* statistical inference, namely, the choice of particular values for α, β, and n. Mathematical theorems proving the existence of a best test, a UMP test, and so forth only specify relative values among these parameters. By themselves these relations provide no answer to questions like: What value of α is appropriate in this test of this particular simple hypothesis, H? If rejecting H when it is true is equivalent to not performing the appropriate action in a definite practical context, the consequences of this mistake may be clearly weighed—perhaps even in monetary units. For example, deciding not to breed those plants when in fact the F_2 probability of dominants is .75 represents a clear loss in potential profits to plant growers. A particular choice of α is then set by determining how great a chance of making this mistake one is willing to risk.

Unfortunately, this way of viewing the acceptance and rejection of statistical hypotheses opens up what at least appear to be large gaps in the Neyman-Pearson model. The gains and losses associated with the consequences of possible actions have become crucial to the choice of test parameters. Yet the formal model contains no parameters representing these crucial factors. Their part in determining other test parameters is apparently to be left to scientific intuition. Neyman and Pearson criticized Fisher for leaving too much of the choice of a critical region to unformalized intuition. It now seems that they, or at least Neyman, are open to the same sort of criticism. The person who most successfully exploited this line of objection was Abraham Wald. He succeeded in generalizing the Neyman-Pearson model of testing statistical hypotheses into a full-blown statistical decision theory. But it is questionable whether the program of statistical decision theory, at least as conceived by Wald, should be regarded as the last major step in the development of testing models of inference or as the first step toward modern information models of inference and decision-making. At any rate, there is no question that the debates since the early 1950s have been profoundly colored by Wald's formulation of the problems.

2.3. *Wald's Model of Statistical Decision-Making*

Wald's approach to the theory of statistical inference and decision-making was a synthesis of two previous intellectual achievements.[16] One, of course, was the theory of hypothesis testing which came to him via Fisher and Neyman. The other was the theory of utility and games as developed in John von Neumann and Oskar Morgenstern's monumental *Theory of Games and Economic Behavior*. In order to keep the length of this discussion from getting completely out of hand I will here drop all pretense of presenting a historical development and simply sketch those parts of utility theory and game theory that are necessary to understand later criticisms of the testing paradigm.[17]

Utility Theory

Utility theory is based on the single primitive notion of preference for one object over another, where the "objects" in question might be things, events, states of affairs, possible worlds, or whatever. I will use the expressions $O_1 > O_2$ to express strict preference of O_1 over O_2; $O_1 \geqslant O_2$ to express weak preference, that is, preference or indifference; and $O_1 = O_2$ to express indifference. Either strict or weak preference may be taken as the basic primitive, with the other two relations defined in obvious ways. The basic theory of utility may then be set out in four innocuous-looking but powerful axioms, with one intervening definition.

Axiom 1. (Closure). For any two outcomes O_1 and O_2, either $O_1 > O_2$, $O_2 > O_1$, or $O_1 = O_2$.

Axiom 2. (Transitivity of weak preference). If $O_1 \geqslant O_2$ and $O_2 \geqslant O_3$, then $O_1 \geqslant O_3$.

Definition. $O = (O_1, O_2, p)$ is a *mixture* of O_1 and O_2 iff O is either O_1, with probability p, or O_2, with probability $(1 - p)$.

Axiom 3. If $O_1 > O_2 > O_3$, then

a. there is some $O' = (O_1,O_2,p')$, where $0 < p' < 1$, such that $O' > O_2$.

b. there is some $O'' = (O_1,O_3,p'')$, where $0 < p'' < 1$, such that $O_2 > O''$.

Axiom 4. If $O_1 > O_2$, then for any O_3, and any p, $0 \leqslant p \leqslant 1$, $(O_1,O_3,p) > (O_2,O_3,p)$.

The basic theorem connecting preference with utility follows:

Theorem. If A_1 through A_4, then there is a real-valued function $U(O)$ such that,

a. $U(O_1) > U(O_2)$ iff $O_1 > O_2$

b. If $O = (O_1, O_2, p)$, then $U(O) = p\,U(O_1) + (1 - p)\,U(O_2)$.

I will not comment here on utility theory as such except to note the utterly crucial part played by mixtures of objects in axioms 3 and 4. The existence of these and similar mixtures will emerge as a major source of controversy later on.

Game Theory

For present purposes it is sufficient to consider only two-person, zero-sum games. Imagine a game with two players, A and B, each having only one play. If A has m options and B has n, there are $m \times n$ possible outcomes, each consisting of a pair of actions, one by each player. These outcomes are the objects of preference and thus the domain of a utility function. Consider the 3×3 game in figure 3, in which the entries in the matrix represent utilities for A. The main assumptions governing the game are these:

1. *Independence.* Each player's choice is independent of the choice of the other.
2. *Perfect knowledge.* Each player knows the complete structure of the game, including all utilities.
3. *Perfect competition.* The utility of any outcome for one player is the negation of the other's utility for that outcome.

It is the last of these that is reflected in the description "zero-sum game."

	B_1	B_2	B_3
A_1	10	−5	−8
A_2	−8	−5	0
A_3	−3	−4	−3

Figure 3

The following definition is crucial:

Definition. U_0 is the *security level* of action A_i iff U_0 is the lowest value of all outcomes possible with action A_i.

In short, the security level of any action is the least one can gain, or the most one can lose, if he chooses that action. Now action A_3 can be seen to have the following two optimal characteristics:

1. It maximizes A's security level (minimax property).
2. It is the best choice if B also maximizes his security level, that is, by choosing B_2.

If these conditions are satisfied, the action A_3 is said to be an equilibrium strategy for player A. The pair of actions (A_3, B_2) is called an equilibrium pair. But not all games yield an equilibrium pair. For example, the 2×2 game in figure 4 has A_2 as the minimax choice for A, but A would do better with A_1 if B takes his minimax choice, B_1. But then B should take B_2, in which case A should take A_2, and so on round and round.

By analogy with mixtures of objects in preference theory, one can define mixed strategies of the form:

$$M_A = (A_1 \text{ with } p_1, A_2 \text{ with } p_2, \ldots, A_m \text{ with } P_m),$$

where $\Sigma_i p_i = 1$. Thus, for any action B_i of B, the *expected utility* of M_A is

$$EU_i\ (M_A) = p_1 U_{1i} + p_2 U_{2i} + , , , + p_m U_{mi}.$$

Now choose that set of probabilities which gives the maximum expected value that can be guaranteed for A regardless of what B does. Thus, we have a natural extension of the concept of a security level for mixed strategies. Similarly, there is a set of probabilities, $q_1, \ldots, q_n$, that provides a security level for player B. If each player chooses the appropriate mixed strategy, a new kind of equilibrium is achieved. This result is summarized in a famous theorem:

Minimax theorem. For any zero-sum, two-person game, there exists an equilibrium strategy (which may be a mixed strategy).

For the 2×2 matrix (figure 4), the probabilities $p_1 = 1/2$ and $q_1 = 3/4$ produce an equilibrium in which the players' security levels are 2.5 and –2.5 respectively.

	B_1	B_2
A_1	3	1
A_2	2	4

Figure 4

Statistical Decision Theory

Let us now recast the problem of testing a statistical hypothesis in an explicitly decision-theoretic framework. Again suppose we are considering two Mendelian models in which the probability of F_2 dominants is .5 and .75 respectively. The investigator has the options of accepting H: $p = .5$ or K: $p = .75$. Taking Neyman at his literal word, "accepting H" would not be a case of "inductive behavior." But breeding or not breeding a certain species of plant would. So the game pits the investigator or plant breeder against nature. The breeder has two possible "moves," breed these plants or not. Nature has two "moves," H or K. Assuming it is profitable to breed only if $p = .75$, U_{21} represents the loss if one makes a Type I error, and U_{12} the loss involved in making a Type II error. Similarly, U_{11} and U_{22} represent the gains in making the corresponding correct decisions. In a concrete, practical, decision-making context of this kind, it is often possible to estimate the potential profits and losses involved for each possible outcome.

The representation of the decision problem in figure 5 makes no reference to any statistical data. Introducing data complicates but does not alter the fundamental structure of the problem. Suppose we breed, independently, n F_2 individuals. The possible results of this experiment are the $n + 1$ possible numbers of dominants; $S = \{0, 1, 2, \ldots, n\}$. In

	B_1 H: $p = .5$	B_2 K: $p = .75$
A_1 Don't breed	$U_{11}(8)$	$U_{12}(-1)$
A_2 Breed	$U_{21}(-3)$	$U_{22}(10)$

Figure 5

Wald's terminology, there are then 2^{n+1} different complete *decision functions*, d_i, that is, functions which for each possible result tell us whether to choose A_1 or A_2. Knowing the probabilities of the possible results given either H or K, and knowing the value matrix (figure 5) for right and wrong choices, we can easily calculate the expected utility of each decision function if H is true or if K is true. Thus, we can construct a new decision matrix in which the choices are not among actions directly, but among decision functions. Once a decision function is chosen, however, an action is selected via the decision function as soon as the result of the experiment is known. The new decision matrix, then, is shown in figure 6.

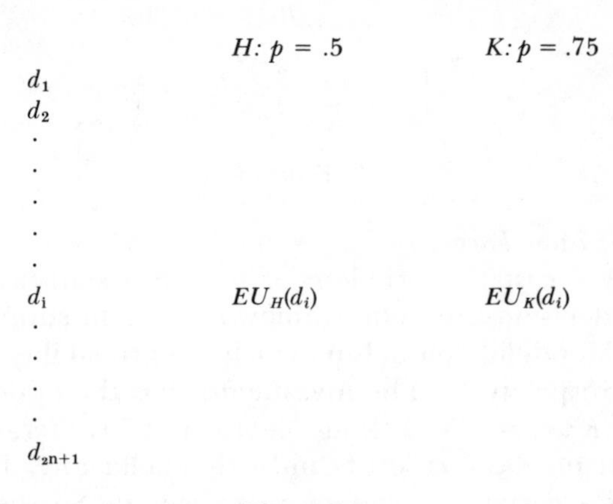

Figure 6

Now whether we consider the original matrix (figure 5) or the more complicated decision-function matrix (figure 6), the fundamental problem is the same: Find, and justify, a rule that picks out the optimal (in an appropriate sense) action or decision rule as a function of the corresponding utility matrix. This may fairly be called *the* fundamental problem of decision-making under uncertainty. Moreover, it remains unsolved, and perhaps unsolvable, for want of a sufficiently general and rationally justifiable definition of an optimal action. Yet several important results have been achieved. One of these is summed up in the following definitions and theorem:

Definition. Action A_i *dominates* action A_j iff $U_{ik} > U_{jk}$ for some k and $U_{ik} \geqslant U_{jk}$ for all other k.

Informally, A_i dominates A_j iff A_i is better than A_j when one hypothesis is true, and at least as good if any other hypothesis is true.

Definition. A_i is an *admissible* action iff A_i is not dominated by any other possible action.

Now consider the possibility that there is a prior probability distribution over the possible states or hypotheses (H and K in the example).

Definition. Action A_i is the *Bayes strategy* relative to prior probability distribution P iff A_i has the greatest expected value relative to P.

(The expected value of A_1 in the original matrix [figure 5] would be $EU(A_1) = P(H)U_{11} + P(K)U_{12}$.) Finally, the following theorem should become obvious after a little thought:

Theorem. The set of admissible strategies is equivalent to the set of possible Bayes strategies.

The set of possible Bayes strategies, of course, is determined by the possible prior probability distributions.

The above theorem and related results were proved by Wald long before Savage made popular the use of subjective prior probabilities. Thus, it was clear all along that a prior probability distribution eliminates the problem of decision-making under uncertainty—because it eliminates part of the uncertainty. With a prior probability distribution the problem is one of decision-making with known risk, and Bayes' rule is generally acknowledged to be correct.

Bayes' rule. Choose the Bayes strategy relative to the given prior probability distribution.

But Wald, being in the tradition of Fisher and Neyman, refused to think of prior probability distributions as anything more than purely mathematical devices. For him, the only probabilities relevant to actual decision-making are the relative frequencies of outcomes of experiments. In most problems the probabilities of hypotheses are at best unknown, if not simply nonexistent.

Rejecting prior probabilities leaves one with two undisputed, but insufficient, rules:

Rule of admissibility. Choose only admissible actions.

Rule of dominance. Choose the action that dominates all others, if one exists.

For most problems there is no dominant action and one is left with several admissible possibilities. In the original plant-breeding matrix (figure 5), for example, the suggested values make both actions admissible. The only other principle that has received widespread support is minimax.

Minimax rule. Choose the action with the greatest security level.

In game theory this rule is justified by the existence of equilibrium pairs. But in decision theory there is no such thing. In spite of the fact that some theorists have written about "games against nature,"[18] one cannot take literally the idea that nature is choosing actions, whether pure or mixed. The only real justification for the minimax rule, therefore, is that it yields the choice with the best guaranteed minimum payoff. It is difficult, however, to regard this choice as uniquely optimal for all decision makers in a context of uncertainty. By the minimax rule it would always be irrational to chance a lower security level for the possibility of a greater maximum payoff. In the matrix shown in figure 7, for example, the minimax rule recommends A_2, while most "rational" people would take A_1.

	S_1	S_2
A_1	0	10
A_2	1	1

Figure 7

One final approach to the fundamental problem should be mentioned—the axiomatic approach. Defining a decision problem in terms of sets of actions, states, utilities, and so forth, one may attempt further characterization using additional axioms on structures of this type. In addition to gaining precise insights into decision problems, one would hope to find a set of axioms sufficient to yield a unique decision rule. Unfortunately, the most plausible set of axioms yields only the following rule:

Average utility rule. Choose the action with the greatest average utility.

But this is equivalent to assuming equal prior probabilities for each state (hypothesis).[19] After nearly two hundred years we are back to Laplace's information model!

With the failure to solve the problem of decision-making under uncertainty, many students of scientific inference and decision-making came to regard all testing models with suspicion. Indeed, several prominent theorists have argued that the only way to complete Wald's program is to bring inference and decision-making under the information paradigm by reintroducing probabilities of hypotheses. I will proceed to examine some of these arguments and then suggest a way of reviving the testing paradigm, at least for inference in scientific contexts.

3. MIXED TESTS AND EMPIRICAL PROBABILITY

3.1. Mixed Tests and Relative Frequencies

My primary concern is to counter certain arguments of Dennis Lindley and L. J. Savage to the effect that the testing models of Neyman and Wald are inherently incomplete, or even contradictory, unless supplemented with prior probabilities of hypotheses. Before looking at these arguments, however, I want to examine another class of reputed objections to Neyman-Pearson models of statistical hypothesis testing. The answer to these latter objections will provide part of the solution to the more serious objections of Lindley and Savage.

Imagine two varieties of a particular plant such that the recessives of both have the same mean height, μ_o, and the dominants have the same mean height $\mu' > \mu_o$. All the heights are normally distributed. The only difference is that the variance of the heights in the first variety is much larger than that of the second, being roughly equal to the difference $\mu' - \mu_0$. Now suppose we have a pure strain of each variety, the only question in each case being whether the strains are pure dominants or pure recessives. Although this is not too realistic scientifically, imagine that for some reason we can only test one of the two varieties, for example, we are pressed for time, equipment, or the like. Finally, just to make things simple, suppose we take only one observation, *x*, the height of one plant. The four distributions relevant to the test are indicated in figure 8.

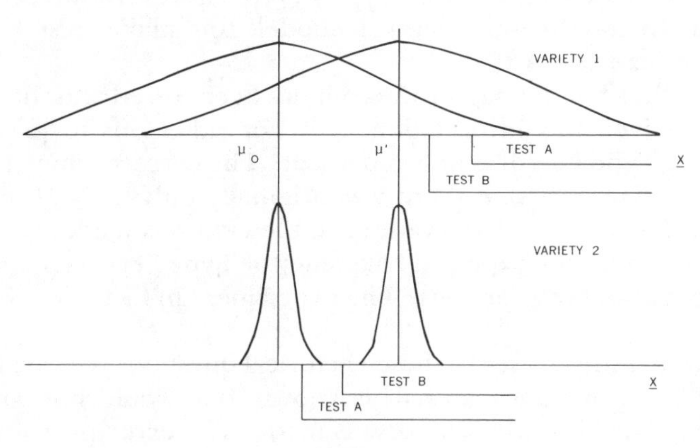

Figure 8

Suppose we decide which variety to test by tossing a fair coin. No matter which variety we end up testing, we have a simple hypothesis *H*: $\mu = \mu_0$ versus a simple alternative *K*: $\mu = \mu'$. Fixing the significance level at .05, there is in either case, by the Neyman-Pearson lemma, a best test of *H* versus *K*. The respective rejection regions for these best tests are:

Test A

$x > 1.64\sigma_1$ if *x* is the height of a plant of variety 1.
$x > 1.64\sigma_2$ if *x* is the height of a plant of variety 2.

In either case the significance level is .05. However, because the variance in heights for variety 2 is much smaller, the power against *K*, that is,

$P_K(X \epsilon R)$, is correspondingly much greater when variety 2 is being tested. This fact, combined with the fact that the variety to be tested is chosen by a coin toss, makes it possible to construct a *mixed* test with the same significance level but greater power than test A regarded as itself a mixed test. Indeed, there is a best 5 percent mixed test defined by the following rejection region:

Test B

$x > 1.28\sigma_1$ if x is the height of a plant of variety 1.
$x > 5\sigma_2$ if x is the height of a plant of variety 2.

As can be seen from the distributions sketched in figure 7, test B has greater power when variety 1 is tested, and only negligibly less power if variety 2 is examined. The significance level, however, is 10 percent when variety 1 is measured. On the other hand, this is offset by the fact that there is only a negligible probability of Type I error if variety 2 is checked, which would be half the time on the average. Thus, if we imagine a sequence of experiments, including the coin toss, test B will produce the same frequency of Type I errors as test A, but fewer Type II errors. In the Neyman-Pearson model, this makes test B a better mixed test than test A.[20]

On the other hand, imagine the coin has been tossed and that it favors variety 1. Given this information, is it not ridiculous to pretend our chance of a Type I error is only 5 percent? The correct value, it seems, is 10 percent. The fact that there was originally only a 50–50 chance of getting variety 1 seems irrelevant once the choice is made. Why should one even consider the expected frequency of Type I errors in a sequence involving tests of both varieties? The experiment in fact performed is on variety 1 only.

I take it that the answer to these rhetorical questions is clear. Once the choice of experimental materials is known, one would not, and *should* not, worry about how this choice was made. The error probabilities are to be determined only by the conditions of the experiment actually performed. The interesting question is whether one can justify this judgment without abandoning essential features of the Neyman-Pearson model.

The basic problem may be restated as follows: Does the Neyman-Pearson model for testing statistical hypotheses provide any basis first for distinguishing a "pure" from a "mixed" experiment, and then for favoring the pure? When the problem is formulated this way, it becomes clear immediately that the Neyman-Pearson testing model, as such, contains no means even for distinguishing pure from mixed tests, let alone ruling out mixed tests as somehow illegitimate. The general structure of a Neyman-Pearson test, as given in section 2.2, is simply that of a set of admissible statistical hypotheses for a given sample space, both partitioned into proper subsets. Additional distinctions are then made in terms of the simple or composite nature of the subsets of the admissible set of hypotheses. There is nothing in any of this that will by itself discriminate a mixed from a pure test. The continuing impact of the above type of example on statisticians and philosophers stems, I think, from the intuitive realization that this is so.

There are several possible options open to one who would like to rule out mixed tests, like test B, and still preserve the basic structure of the Neyman-Pearson model. The most obvious is to look for a formal requirement that could be built into the model itself. The trick would be to find some restriction strong enough to eliminate the intuitively undesirable mixed tests without changing fundamental features of the model. It would be very difficult to argue that this could not be done, so I will attempt no such argument here. Nor will I survey various suggestions that have been made. Instead I will suggest another approach altogether.

Any formal structure may have unintended models. For example, a galaxy provides the elements for a model of finite probability spaces, with the relative masses of the stars providing the basic probability measure. Yet no one takes this as a reason for inquiring into the "probability" of some set of stars or, worse yet, for rejecting the theory of finite probability spaces. Similarly, the formal structure of the Neyman-Pearson account of statistical tests may have unintended models. In particular, tests like test B are, I will argue, unintended models of the Neyman-Pearson formalism. Since no formal conditions can rule out the possibility of there being unintended models, one is forced back upon informal considerations. The problem is to be sufficiently systematic to avoid mere *ad hoc* pronouncements. In the present case, the basis for a systematic account can be found in the physical models of probability that provide the backbone for any application of the Neyman-Pearson formalism.

As noted earlier, the development of testing models of inference since the mid-nineteenth century has been accompanied by the identification of empirical probabilities with limiting relative frequencies. So strong is this association that statisticians in the Fisher-Neyman-Wald tradition are commonly called "frequentists" in much the way that Bayesian theorists are called "subjectivists" or "personalists." It is just this identification of probability with relative frequency that I wish to challenge.

The trouble with frequency interpretations of probability, at least as concerns the present problem, is that they are too inclusive. Standard frequency interpretations provide no basis for discriminating against the hypothetical infinite sequence in which the error probabilities of test B are interpreted as limiting relative frequencies. One might, of course, attempt to devise and justify restrictions on the frequency models themselves. Recent work on "the problem of the single case" and "the problem of the reference class" may be viewed in this way.[21] Being skeptical of the success of these attempts to refine frequency models, I will not pursue this line of attack. Instead I will try to show how a propensity interpretation of probability might provide the needed restrictions—and in a systematic way.

The number of variations on a propensity interpretation of probability is nearly equal to the number of authors advocating such interpretations.[22] For present purposes, fortunately, the similarities are more important than the differences. Almost any propensity interpretation will yield the kind of restrictions we seek. I will distinguish only two types of propensities, "relative" and "absolute."

3.2. Relative and Absolute Propensities

Relative frequency models of probability structures make probability an attribute of *sequences* of objects, events, or whatever. Thus $P(A) = r$ will be true in a frequency model in which the limiting relative frequency of objects with property A in an infinite sequence of such objects is r. The fundamental difference between frequency models and almost any propensity model is that propensities are attributes not of sequences but of physical systems. The problem has been to say more about what kind of an attribute a propensity might be. The most common approach is to identify propensities with dispositions of chance setups to produce particular distributions of relative frequencies upon repeated trials. The measure of the propensity of a single outcome in a finite set, for example, would then be the relative frequency with which that outcome would be produced were one to repeat trials indefinitely. This, in general terms, seems to be the view both of philosophers like Popper, Hacking, and Isaac Levi, and of theoretical statisticians who follow Harald Cramer. The main difficulty with this view, of course, is understanding counterfactual assertions about what relative frequencies would occur were some process to be repeated indefinitely. Some writers like Hacking, Levi, and even much earlier, R. B. Braithwaite, have explicated the meaningfulness of propensity hypotheses in terms of rules for the assertability of such hypotheses. In spite of its long empiricist and pragmatist pedigree, this approach seems to me fundamentally mistaken. It is precisely the job of a testing model of statistical inference to provide rules for accepting and rejecting statistical hypotheses, that is, hypotheses about propensity distributions. But such models presuppose an understanding of what it is that is being accepted or rejected. Thus, what we want for propensity hypotheses are truth conditions at least.

In recent years the most promising approach to an analysis of counterfactuals has been via the sorts of possible worlds semantics developed for modal logics.[23] It is natural to attempt some such analysis of propensity statements understood as counterfactual, or at least subjunctive, assertions about limiting relative frequencies. Henry Kyburg has recently sketched such an analysis using sequences of possible worlds.[24] Such analyses are bound to increase the precision of future discussions of propensities. They do provide honest-to-goodness truth conditions for propensity hypotheses. Unfortunately, like possible world analyses of regular counterfactuals, Kyburg's analysis is still very formal and abstract. In particular, it provides no more basis for distinguishing mixed tests A and B than does any frequency interpretation. If we are to legitimate our intuitions regarding these tests, we must seek something which, though perhaps less formally precise, does contain the needed emprical content.

The key to a deeper understanding of propensities is, I think, the distinction between deterministic and indeterministic systems. While this distinction is itself in need of much analysis, it is already clear enough to provide some appreciation for two radically different accounts of physical propensities. Let us simply say that a deterministic system is one whose state at any given time is uniquely determined by its state at any earlier time. The states of an indeterministic system are not so determined. Now, since at least the beginning of the nineteenth century, it had been explicitly thought that all physical systems are deterministic, but that owing to our ignorance of many relevant factors, systems that seem to be in the same initial state end up in different final states. It is, then, an interesting empirical discovery about some types of systems that, upon repeated trials, the different outcomes occur with fairly stable relative frequencies. Ignoring all the problems of inductive inference, we do discover rough-and-ready statistical laws, for example, that the frequency of dominants in the F_2 generation is roughly 3/4. But this law is relative to a description based on our knowledge of the process in question. This point is easily missed because the appropriate description is usually quite obvious, as with plant-breeding or coin-tossing. Yet the relativization to a particular description is implicit in our reference to repeated trials. Strictly speaking, no trial is repeated in all details; what is repeated is only a trial of the same, necessarily partial, description. If we could observe the details of gamete formation in a particular egg, for example, we might discover a refined description such that 90 percent of all F_2's answering this new description would be dominants. This would be a new statistical law relative to a new description. A still more refined description might pick out all and only the dominants, thus yielding a universal causal law.

As Lawrence Sklar has argued, if one identifies propensity hypotheses with the above sort of rough-and-ready statistical laws, then propensities may be fully analyzed in terms of the operation, in designated types of systems, of both known and unknown causal laws with both known and unknown initial conditions.[25] The distributions of frequencies we observe in trials answering the same description are the product of the distributions of these unknown laws and initial conditions. Insofar as counterfactuals or modalities are involved, they are the standard kind related to standard causal laws. In short, in deterministic contexts, "propensity" is merely an abbreviation for a particular arrangement of causal laws and produced frequencies. A propensity is not an essentially new kind of property.

The situation is radically different if one takes seriously the possibility of ontologically indeterministic systems, that is, systems for which deterministic laws not only are not known but do not exist! In principle there are two classes of indeterministic systems, those in which there is no connection whatsoever between earlier and later states, and those for which an earlier state determines a probability distribution over later states. The only plausible examples of indeterministic systems, that is, quantum systems, are probabilistic. The problem is that interpretations of quantum theory are even more controversial than interpretations of probability. Nevertheless, it is not implausible to regard individual quantum systems, for example, a single radioactive nucleus, as possessing propensity distributions over later possible states, for example, decay in less than one minute or decay in a minute or more. Such a propensity would be a kind of weak, but weighted, causal force. It would be absolute in the clear sense that it is a characteristic of a particular trial of a designated system—no reference to a repeatable *kind* of system is necessary. Moreover, the lesson of von Neumann's and later proofs is that if the current theory is correct, there are no hidden variables that determine a unique outcome for any particular trial. There is no more refined description to be discovered.[26]

For absolute, "single-case" propensities, there is no "problem of the single case" or "problem of the reference class" of the type that has vexed frequency theorists from Venn to Hans Reichenbach and Wesley Salmon.[27] Absolute propensities are by nature characteristics of individual trials of particular systems. On the other hand, such a conception of physical probability is too rich for the empiricist metaphysics and epistemology that have dominated philosophy of science for the past several generations. It is not compatible with a Humean analysis of causality. Nor is it compatible with a foundationist epistemology that would make singular facts the sole basis for all knowledge.[28] It is not surprising, therefore, that most propensity theorists have desired a conception of propensities that is richer than that of relativized propensities but less rich than that of absolute propensities. Whether this is possible and whether one or another or both conceptions are necessary for a complete account of scientific inference are fortunately questions we do not have to answer here. Even relativized propensities are sufficient to support the rejection of some mixed tests as inappropriate in standard scientific contexts.

3.3. Mixed Tests and Propensities

As we have seen, interpreting the error probabilities in Neyman-Pearson tests as mere relative frequencies provided no basis for discriminating against mixed tests like test B. Understanding error probabilities as propensities, even just relativized propensities, does provide the needed basis for discrimination simply by forcing us to consider specific causal processes, or kinds of processes, rather than just sequences. In order to apply the Neyman-Pearson formalism we must first pick out the causal process whose propensities are in question. This means we can use criteria for selection appropriate to causal processes, criteria which may differ from those appropriate to the selection of sequences.

In the example described earlier, there are three processes to consider: the generation of offspring by the first variety, generation by the second variety, and a coin toss followed by generation by one variety or the other depending on the outcome of the toss. Now we know that the propensities for both types of error can be made better for the mixed process—test B. But this is only on the assumption that our concern is merely with the overall relative frequency of mistakes regardless of which variety we test, an assumption forced on us by the frequency interpretation of error probabilities. Rejecting frequencies in favor of propensities allows us to reject this assumption in favor of our scientific common sense, which tells us that our concern is with the probability of mistakes regarding specific causal processes of particular scientific interest. Generation of offspring by variety 1 and generation by variety 2 are processes of scientific interest. Generation by one or the other depending on the flip of a coin is not. Even in the unlikely situation of being able to test only one of the two varieties, our scientific concern is with the variety tested, not with the overall process which includes the process by which we select which variety gets tested. The error probabilities of interest, therefore, are those for one set of admissible hypotheses or the other, not a mixture of the two.

A propensity interpretation of physical probabilities allows us to reject some applications of the Neyman-Pearson formalism, that is, some mixed tests, as scientifically inappropriate, but it does not rule out all possible uses of mixed tests. Suppose, for example, that the tests for dominants and recessives in the two plant varieties are part of a quality-control setup in a plant-breeding establishment. At fixed intervals, one variety or the other is selected for testing by some binomial process with equal probabilities. In this sort of context one may well be more concerned with the overall probabilities of errors than with the errors for each variety separately. Indeed, the testing process may proceed automatically with no one even bothering to check which variety is chosen for testing at any time interval. In this case it would be quite rational simply to build in test B with the outputs being "Accept H_0" or "Reject H_0," with 'H_0' referring ambiguously to the recessives of either variety. It is difficult, on the other hand, to imagine a scientific research situation in which one would not care to know which variety is being tested. This is why test B seems so counterintuitive.

It is interesting to note that Fisher's later writings are full of polemics against those who would assimilate methods of scientific inference to techniques for industrial quality control. In this, as in other matters, Fisher's intuitions were better than his conceptual or mathematical machinery could justify. Replacing the frequency interpretation of probability with some form of propensity interpretation provides some conceptual basis for these intuitions. But much more would be needed if anything like Fisher's conception of scientific inference were to be adequately vindicated.

4. ARE NEYMAN-PEARSON TESTING MODELS INCOMPLETE OR INCOHERENT: THE ARGUMENTS OF LINDLEY AND SAVAGE

Wald's brilliant formulation of the problem of decision-making under uncertainty was greeted with great intellectual enthusiasm. But the enthusiasm was soon dissipated when no equally compelling solution was forthcoming. Outside explicit game-theoretic contexts, the arguments for minimax were never generally regarded as sufficiently persuasive, and rightly so. Moreover, no other proposal was ever really in the running. Of course there were interesting new mathematical results *within* the general framework set by Wald. David Blackwell and M. A. Girshick's *Theory of Games and Statistical Decisions,* published in 1954, provides numerous examples of such results. Ultimately, however, L. J. Savage's *Foundations of Statistics,* published the same year, had far greater influence on theoretical statistics.

From a mathematical point of view, Savage's work was very similar to that of his fellow theoretical statisticians. He was seeking an axiomatic characterization of the optimal action in a basic Waldian decision problem. He deviated mathematically in that he explicitly sought axioms which would yield both a utility function over all outcomes and a probability function over all states (or hypotheses). The optimal action is, then, the Bayes strategy relative to the given probability function. Savage's main deviation from his Waldian contemporaries, however, was more philosophical than mathematical. He proposed that we explicitly regard the utility and probability functions as representing the *subjective* values and beliefs of an ideally rational agent. By turning his back on the century-old tradition which identified probability with empirical relative frequencies, Savage in one stroke eliminated the problem of decision-making under uncertainty and created a unified, mathematically sophisticated information model of inference and decision-making.

It is hardly to be expected that anyone would attempt so radical a break from his own intellectual upbringing unless he thought he had very good reasons for rejecting his earlier views. And indeed, both Dennis Lindley and Savage report having discovered such reasons in the early 1950s. It is these arguments I wish to examine here, for if they are sound, one should probably forget about the testing paradigm of statistical inference altogether. I hope to show that the arguments do not show nearly as much as either Lindley or Savage has claimed. Moreover, what they do show can be accommodated within the testing paradigm of scientific inference.

It is, of course, no argument against Bayesian information models that the reasons leading Bayesians gave for rejecting certain testing models were less conclusive than they thought. My purpose, however, is not to criticize information models directly. It is, rather, to argue the viability of the testing paradigm of scientific inference, and thereby indirectly undercut support for the use of information models, at least in scientific, as opposed to practical, contexts.

4.1. Neyman-Pearson Tests and Prior Probabilities: The Lindley-Savage Argument

The Lindley-Savage argument is most easily grasped for the case of testing a simple hypothesis, H, against a simple alternative, K.[29] This is an appropriate point ot attack because here there is a clearly demonstrable best α-level test of H versus K (the fundamental lemma). A convenient example, once again, would be the investigation of competing Mendelian models predicting probabilities for dominants of .5 and .75 respectively.

Any test of a simple hypothesis versus a simple alternative may be represented as a point in a unit square with axes α and β, where:

$$\alpha = P_H(X \epsilon R) = \text{probability of Type I error.}$$
$$\beta = 1 - P_K(X \epsilon R) = \text{probability of Type II error.}$$

The fundamental lemma in effect fixes a value for β given values for α and the sample size, n. It does not, however, determine what the values of n and α should be. It is just this supposed gap in Neyman-Pearson tests that inspired Wald to develop a more general statistical decision theory. From this more general Waldian perspective we see that any α-β pair, (α,β), is associated with a decision rule that says "Accept H" if certain outcomes are observed and "Accept K" if they are not. So picking an α-β pair is choosing a decision rule. The only undisputed principle for making such a choice, as we have seen, is admissibility. Thus, if (α,β) and (α',β') are two tests in a given set of possible tests, and if $\alpha \leq \alpha'$ and

$\beta<\beta'$ or $\alpha<\alpha'$ and $\beta\leqslant\beta'$, then (α,β) dominates (α',β'), which makes (α',β') inadmissible. In short, if one of two tests has lower probabilities for both Type I and Type II error, or lower for one and equal for the other, the other test is inadmissible. But of course there are lots of sets of possible tests, none of which dominates any other. For example, if we fix *n*, the resulting admissible tests lie on a convex curve (like the dotted line in figure 9) between (0,1) and (1,0). The latter two points represent the decision rules "Accept *H* regardless of the data" and "Accept *K* regardless of the data." But where along the curve we should fix α is a question of decision-making under uncertainty for which there is no accepted solution. And this is true even assuming we can attach utilities to the four possible outcomes, including the two types of error. Now what Lindley and Savage claim is that on principles accepted by the Neyman-Wald tradition, the Waldian formulation of the problem by itself already commits one to assuming a value for the ratio $P(H)/P(K)$, the prior probabilities of *H* and *K* respectively.

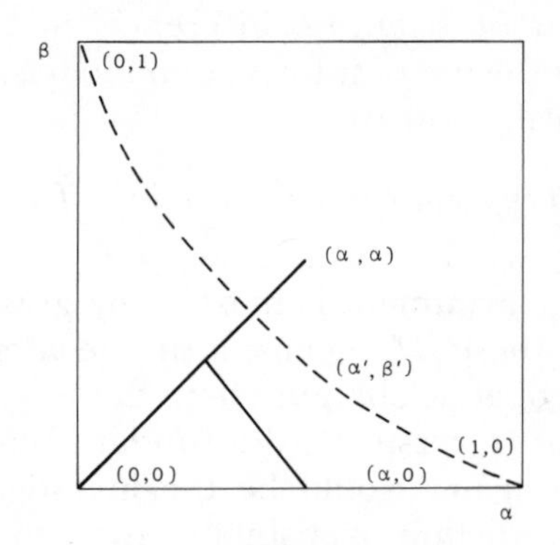

Figure 9

I will proceed to reconstruct the argument semiformally, stating one lemma and one theorem, and sketching proofs. In addition to assumptions already discussed, the first lemma requires the assumption that preference varies continuously. That is, if one moves continuously from a test preferred over a given test to a test less preferred than the given test, there must be a test in the continuum which is indifferent to that given test.

Lemma. If (i) an admissible test is always preferred to an inadmissible one, (ii) a mixture of two tests is a test, and (iii) preference changes continuously, then there are at least two tests between which one is indifferent.

Proof. By admissibility $(0,0) > (\alpha,0)$ and $(\alpha,0) > (\alpha,\alpha)$. The line *b*, with end points (0,0) and (α,α) represents a continuum of mixed tests of the form $k(\alpha,\alpha) + (1\text{-}k)(0,0) = (k\alpha,k\alpha)$, where $0 \leqslant k \leqslant 1$. By the continuity of preference, there must therefore be a number k' such that $(k'\alpha,k'\alpha) = (\alpha,0)$.

Though it is not necessary for the theorem which follows, it is very interesting to note that if there are two indifferent tests, then indifference holds among these tests and all tests along the straight line between. This follows because if one is indifferent between two tests, say (α,β) and (α',β'), one must be indifferent between any mixture of the two, since the mixture is just one or the other with some probability. But the form of a mixture of two tests is $k(\alpha,\beta) + (1-k)(\alpha',\beta')$, where $0 \leqslant k \leqslant 1$, and this is just the straight line between the points (α,β) and (α',β'). What this shows is that indifference curves among tests of simple hypotheses are straight lines. This is worth noting because it is tempting to regard the fixed sample size convex set of admissible tests (dotted curve) as an indifferent set. The elementary considerations just stated show that this would be a mistake. The heart of the Lindley-Savage argument is contained in the following theorem:

Theorem. Being indifferent between two tests $t = (\alpha,\beta)$ and $t' = (\alpha',\beta')$ is equivalent to holding H and K with prior probabilities in the ratio

$$\frac{P(H)}{P(K)} = - U_{II}/U_{I} \text{ slope } (t,t')$$

where U_I and U_{II} represent the utilities (or disutilities) of making a Type I or Type II error respectively.

Proof. Assume that indifference between t and t' implies that the expected value of the two tests is the same.

$$EU(t) = P(H) \times \alpha \times U_I + P(K) \times \beta \times U_{II}.$$

$$EU(t') = P(H) \times \alpha' \times U_I + P(K) \times \beta' \times U_{II}.$$

Setting these equal yields the desired result. I will reserve comment on this result until I have presented Lindley's closely allied arguments.

4.2. Is Neyman-Pearson Testing Incoherent: Lindley's Argument

Neyman and Pearson's fundamental lemma is proved by fixing the significance level, α, and sample size, n, and then minimizing the probability of Type II error. This strategy is followed in statistical practice as well. One sets the significance level at some conventionally agreed value, α, say 5 percent, and then adjusts sample size and power—if one is careful. What Lindley has shown is that if we formalize this practice in a natural rule of preference regarding tests in the α-β plane, we are led into flatly incoherent preferences, that is, simultaneously strictly preferring t to t' and also t' to t.

Assume first that there is a fixed significance level, α_o, such that any test on the line $\alpha = \alpha_o$ is preferred to any test not on that line. Secondly, assume that for any two tests on the α_o line, the one with the lower value of β is preferable. Given these two assumptions, Lindley argues as illustrated in figure 10. Looking first at the line *bac,* we see that neither *a* nor *c* has a lower value for both α and β. Thus neither dominates the other, so both are admissible. But by the first assumption, test *a* is preferred to test *c* because it is on the line $\alpha = \alpha_o$. Looking at line *dae* we conclude similarly that *a* is preferred to *d.* Thus, $a > c$ and $a > d$. Now looking at *c* and *d,* either $d > c$, or $c > d$, or $c = d$. Now if $d > c$, then $d > f$, since *f* is a mixture of *d* and *c;* and if $d > c$, *d* is preferred to any mixture of *d* and *c.* Similarly, if $c > d$, then $c > f$. Finally, if $c = d$, then $c = f = d$. Now since $a > c$ and $a > d$, it follows that $a > f$. But this contradicts the assumption that if α is fixed, the test with lower β is preferable, since this implies $f > a$. So the given assumptions do indeed generate incoherent preferences.[30]

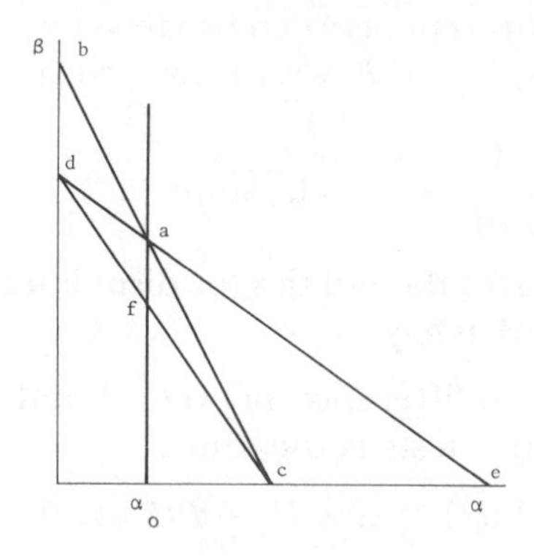

Figure 10

An analogous argument may be given if the preferred tests are determined by the minimax criterion rather than by a fixed $\alpha = \alpha_o$. A minimax test is one for which the expected value of the test is the same whether H or K is true, that is, $EU(H) = EU(K)$. This implies that $\alpha U_{\mathrm{I}} = \beta U_{\mathrm{II}}$, so the preferred tests lie on a line from the origin with slope $U_{\mathrm{I}}/U_{\mathrm{II}}$. Of tests on this line, those closer to the origin are preferable. A similar construction shows that these assumptions also lead to incoherent preferences.

These considerations form the basis of Lindley's oft-repeated claims that the "5 percenters" and "minimaxers" are simply incoherent. Let us now attempt to assess just how deep these arguments cut.

4.3. What Do Lindley and Savage's Arguments Really Prove?

It is obvious that both the Lindley-Savage argument and Lindley's later arguments depend crucially on assuming that a mixture of tests is itself a test. By rejecting this assumption one blocks the arguments at the start. But even if it is right to reject mixed tests, this response is too short. The use of mixed tests was part of Waldian statistical decision theory from the start. Lindley and Savage had every right to assume it. Moreover, the arguments bring out other points of interest. I will return to the issue of mixed tests below.

The most important ingredient of the argument is the realization that indifference curves in the α-β plane must be straight lines. This follows from standard Waldian assumptions and must be admitted by any theoretician in the Neyman-Wald tradition. Anyone who thought otherwise was simply mistaken. I think, however, that Lindley, Savage, and others have overestimated the extent to which their more conventional colleagues, for example, E. L. Lehman,[31] were inclined to make such a mistake. What is true is that if one fixes the sample size one is left with a set of admissible tests which lie, not on a straight line, but on a convex curve (see dotted curve in the α-β diagram, figure 10). If we have no decision principles beyond admissibility, we have no way to select one test from all these admissible tests and are thus in a special sense left indifferent. But this is not indifference in the sense of standard preference theory. If it were, then to the extent that preference theory is normative, one would be forced to claim that we *should* be indifferent among admissible tests. But no one ever seriously held this, because if it were true there would be no point in looking for a criterion that selects one strategy from all admissible strategies—which is just the fundamental problem of decision-making under uncertainty. In sum, by pointing out that indifferent tests lie on a straight line, one is merely drawing attention to something Wald knew all along, namely, that we need some decision principle beyond admissibility.

What about the claim that admitting two indifferent tests commits one to a value for the ratio of prior probabilities $P(H)/P(K)$? The key to this argument is the assumption that indifference between two tests is to be understood as assigning the tests equal expected value. Thus, the prior probabilities are introduced not so much by there being indifferent tests as by the assumption that a test has an expected value. In the Neyman-Wald framework, a test only has expected values relative to H or to K. There is no overall expected value of the test just because there are no prior probabilities. So it is fair to say that Lindley and Savage have simply begged the question for the existence of prior probabilities by interpreting indifference as equality of overall expected value. There is no purely logical reason why indifference of tests must be so construed.

It might be thought that neither Lindley nor Savage ever intended to claim that a ratio of prior probabilities could be generated on the basis of accepted Waldian principles alone. Rather, they may only wish to claim that the choice of an indifference set of tests is "equivalent" to holding prior probabilities in the given ratio.[32] But this claim is misleadingly ambiguous. On the one hand it may simply mean that there is a *mathematical* equivalence. The given probability ratio obtains if and only if two particular tests are indifferent. But this is merely an extension of Wald's own result that every admissible strategy is a Bayes strategy. Surely Lindley and Savage meant to claim more than this. Wald's result involves no more than a claim about a possible mathematical function with the formal properties of a probability function. This is how Wald himself understood it. Lindley and Savage want to interpret this function as representing the beliefs of an ideal agent. This interpretation, however, in no way follows from the mathematical equivalence. To claim equivalence, with the interpretation added, is again begging the issue.

Finally, it might be said that an agent who is indifferent between two tests will behave as if he had prior degrees of belief in the given ratio. This is true, but not much follows. It surely does not follow that the agent in fact has those degrees of belief. It does not even follow that assuming such beliefs is the best way to understand rational actions. Either of these claims requires much more backing than the stated argument provides.

Lindley's arguments require less discussion. If the stated assumptions correctly represent the preferences of a fixed significance level methodology or of minimax, then those preferences are indeed inconsistent. And it is difficult to imagine more plausible interpretations. Apart from introducing subjective prior probabilities, the only course of action is to look for some principles that will pick out a uniquely most appropriate test from among all possible α-β pairs. The question is whether one can do this in a way that by-passes the apparently insoluble problem of decision-making under uncertainty. I will sketch a possible approach in part 5.

I would like to return briefly to the question whether mixed tests are legitimate in scientific contexts. Denying that a mixture of tests must itself count as a legitimate test does block all of Lindley and Savage's arguments. But it leaves unresolved the fundamental problem so forcefully exposed by Lindley's demonstrations. The basis of my objection to mixed tests was that the relevant probabilities for scientific purposes are propensities, whether relativized or absolute, of some physical process. Seldom is a probabilistic disjunction of processes itself a process of any scientific interest.

In the example involving two different varieties of plant, the scientific basis for distinguishing the two processes is clear enough. The case is less clear, however, if we have but one type of organism and the question is whether a given trait is transmitted with probability .5 or .75. Here there is no difference in subject matter, but only in the values for α and β, which typically, but not necessarily, involve a difference in sample size. What, for example, could be the scientific basis for distinguishing two testing processes that differ only in their sample size and power against a simple alternative? The answer must be that the testing processes differ in their scientific interest or value just because of the difference in sample size and power. There is nothing else.

Our situation regarding such mixed tests, therefore, is this: A propensity interpretation, unlike a frequency interpretation, provides a physical (or metaphysical) basis for *distinguishing* one testing process from another. But it does not by itself provide an obvious basis for *preferring* one testing process to another. For this we need additional rules, and better ones than the simple minimax and 5 percent rules that Lindley has so effectively demolished.

5. THE RATIONALITY OF NEYMAN-PEARSON TESTS

5.1. Satisficing and the Methodology of Statistical Hypothesis Testing

I will now sketch a positive account of a testing model of scientific inference. The logical core of this account is the Neyman-Pearson model of statistical hypothesis testing. This core will be supplemented with some methodological principles, including a rough decision strategy known as satisficing. The coherence of the resulting model will be illustrated in connection with tests of both simple causal hypotheses and complex causal theories. Indeed, reference to some such scientific context will be seen to be essential to a full understanding of tests of statistical hypotheses themselves.

Philosophers sometimes distinguish the logic of inductive inference from the methodology used in applying that logic.[33] The need for such a distinction, which is basically that between pure and applied mathematics, seems undeniable. The problems are where and how to draw the line. One strategy for solving these latter problems is to formalize more and more of the total process and hope that the question of how to apply the whole formal structure will eventually become relatively trivial. Nevertheless, since one must at some point jump the gap between the pure formal structure and its application, it is, in any area, an open question how far it is fruitful to push the process of formalization.

Turning to the issue at hand, we have seen that while the Neyman-Pearson model of statistical hypothesis testing provides optimal relations among the test parameters, α, β, and n, it does not prescribe any particular set of parameters. By enriching the formal structure with variables for the utility of correct and mistaken conclusions (now actions), Wald in effect formalized the choice Neyman and Pearson left unspecified. But Wald's general program seems unrealizable in its original form. Its main contribution may turn out to be that it provided a framework ready-made for the reintroduction of probabilities of hypothesis into statistical theory—thus giving up the testing paradigm altogether. My suggestion for avoiding this path—at least for scientific inference—is not to formalize the choice of test parameters into a full-scale decision theory. Rather, I will take the Neyman-Pearson model as the *logic* of testing and provide it with appropriate *methodological* rules for determining test parameters in particular kinds of scientific contexts. This is a more restricted and thus, I hope, easier task than solving the general problem of decision-making under uncertainty.

The unifying principle in my methodology for the application of Neyman-Pearson models is a rough decision strategy known as satisficing.[34] Thus, there remains a general sort of connection between hypothesis testing and decision theory. But satisficing, unlike minimax, is not intended to be a completely general decision strategy. Indeed, it is not entirely clear just what is built into this strategy. For present purposes, the following will suffice: Imagine a standard decision matrix with several actions and states. Look for an action where the utility payoff is at least acceptable for every state. If there is such an action, take it. Of course if there is an action with higher values for some states, take the better action. But if there is no such satisfactory action, you are out of luck. Change the problem, or find some other decision strategy. Thus, unlike minimax or other general decision strategies which yield an optimal action for any matrix, satisficing may yield no decision at all. There may be no action that is good enough.

Now consider again the example of testing simple H versus simple K. Here there are two possible states. A satisfactory decision rule would be one for which the expected disutility of either a mistaken acceptance or a mistaken rejection is just acceptable. Given values for the disutilities of mistakes, this puts an upper bound on acceptable values of α and β. Moreover, if we can assume, as seems reasonable, that there are no trade-offs between these upper bounds and the cost of sampling, then we have no reason to prefer any values other than these upper bounds. We have given up trying to find a decision strategy that would pick out an optimum α-β-n set. Such a strategy would be necessary if we wanted to buy lower error probabilities at the cost of larger samples. Thus, everything hinges on our being able to determine minimally acceptable disutilities for both types of error, or, alternatively, the largest acceptable error probabilities. How might this be done?

Neyman introduced the idea of determining the disutility of errors by identifying the acceptance of hypotheses with the choice of a particular practical action. In so doing he took the first major step toward a full-fledged decision theory. I wish to go back to the notion that accepting (or rejecting) a hypothesis is an independent kind of action. In order to determine the utilities of such actions it will be necessary to say what consequences they have in scientific contexts. And here the restriction to scientific contexts is essential. To accept a hypothesis is to take it to be true. But there must be some restrictions on this taking to be true. It is possible that a biologist in the course of his research be perfectly justified in accepting the statistical hypothesis that the probability of dominants in a breeding experiment is greater than .75. Yet a plant breeder whose total personal fortune hangs on a related decision might be irrational in accepting the same hypothesis on the same evidence. By restricting the context in which acceptance and rejection of hypotheses takes place, I hope to avoid just this sort of problem.[35] This leaves us with the problem of characterizing the restrictions for specific kinds of scientific contexts.

Too much philosophical writing on scientific inference takes place at such a high level of generality and abstraction that no distinctions among kinds of scientific inquiries are relevant. Yet one would think that some of the many possible ways of categorizing scientific activity would have epistemological significance. For present purposes I will introduce a simple developmental distinction between two stages of inquiry. This is still highly idealized, but it at least makes it possible to illustrate how different kinds of contexts could require different parameters for tests of the same hypothesis.

5.2. Testing Simple Causal Hypotheses: Exploratory Inquiry

The most significant occurrence in the development of any scientific field of inquiry is the formulation of an explicit, comprehensive, and unified theory of the subject. Natural selection, Mendelian genetics, and molecular genetics have provided such turning points for much of biology. This is not to say that research prior to such events is not guided by theoretical ideas. But these are vague prototheories, and the purpose of experiments at this time is more to develop a precise theory than to test one. No real test of a theory is possible until it is sufficiently well articulated to have clear experimental consequences. Inquiry prior to this time, which I will call "exploratory inquiry," seems primarily directed at establishing simple causal hypotheses of the form: In systems of type *S*, *C* is a causal factor (positive or negative) for the occurrence of *E*. It is such hypotheses, and not simply data, which form the experimental base to be explained by any proposed theory. As an example, one might imagine radiation experiments on animals in which the effect is to decrease the viability of recessives.

The analysis of simple causal hypotheses is itself a controversial topic, but it need not be so here. Any account of how simple causal hypotheses are tested should be compatible with many different analyses. Since propensities have already been introduced, it is convenient to analyze simple causal hypotheses in terms of propensities. I will say merely that *C* is a causal factor (positive or negative) for *E* in systems of type *S* iff adding *C* to systems of type *S* changes (increases or decreases) the propensity of *E*. On this analysis, the difference between simple causal hypotheses and propensity hypotheses is just scope. A propensity hypothesis applies to a single system or a finite set of systems involved in a particular experiment. The corresponding causal hypothesis applies to all systems of a particular type. If, as would usually be the case, the propensities are relativized to a particular description, a further analysis of the underlying deterministic causes is needed. There is, however, no need to pursue such an analysis here. Nor need we concern ourselves with so-called spurious correlations, though this is an interesting topic.

There is obviously an inductive gap between the statistical (propensity) hypothesis that might be accepted through an appropriate experiment and the corresponding simple causal hypothesis. Scientists pay homage to this gap, though more in word than in deed, by recognizing the need to replicate experiments in different laboratories and so forth. Apart from this proviso, the step from accepted statistical hypothesis to accepted causal hypothesis is usually made quite automatically. Henceforth I shall do the same.

In order to assess the disutility of mistakenly accepting or rejecting a simple causal hypothesis we must know what role such hypotheses play in exploratory inquiry. One role is immediately clear. An accepted simple causal hypothesis is regarded as a *prima facie* datum which any proposed theory will have to explain, or at least not contradict. It is only a *prima facie* datum, of course, because no proposed theory could be expected to agree with all previously accepted causal hypotheses, if only because it is likely that some of these hypotheses were mistakenly accepted in the first place. Nevertheless, accepted causal hypotheses do put strong constraints on what theories will be regarded as serious candidates for later independent tests.

A second role for accepted causal hypotheses is as assumptions to be taken as given in designing tests of other simple causal hypotheses. For example, suppose we find that radiation increases the probability of dominant offspring (presumably by lowering the viability of the recessives). The first time this is tested, hypotheses asserting a lower than normal probability for dominants might be admissible components of a composite alternative. In later tests these alternatives would no longer be regarded as admissible. Many questions might be raised about the epistemological status of sets of accepted hypotheses generated in this way, but I will not pursue them here.

Summing up thus far, to accept a simple causal hypothesis in exploratory inquiry is to regard it as true for the purposes of designing later tests of other causal hypotheses and for developing and judging new theories of the kind of system in question. I will assume that these are the main roles played by accepted hypotheses in exploratory inquiry. The nature of the disutility attaching to mistaken acceptances and rejections follows immediately. Suppose a false simple causal hypothesis is accepted as true. In classic experimental designs this would be associated with the mistaken rejection of a null statistical hypothesis. First, the mistake is likely to be compounded if this hypothesis is assumed in the design of later tests. The net result is likely to be contradictions within the set of accepted hypotheses. This can be helpful insofar as it indicates that something has gone wrong, but it is very troublesome to remedy. Second, theorists will be engaged in trying to accommodate their fledgling theories to a nonexistent phenomenon. This could lead the whole field down a long blind alley.

Now consider the mistaken rejection of a true causal hypothesis, which commonly corresponds with the mistaken nonrejection of a null statistical hypothesis. Though logically no different, such mistakes would seem to be generally less serious. First, a null conclusion generally will not stimulate further experiments in which it would be assumed true. So a null conclusion tends not to compound itself in future experiments. Second, since many more things are not importantly related than are, a new null result is not usually a stimulus to theoretical speculation—at least in exploratory inquiry. A null result in response to a non-null prediction by theory is an important scientific event, but this occurs at a later stage of inquiry.

The goal of exploratory inquiry is a well-articulated theory that would merit testing on its own. The utility and disutility of correct and mistaken acceptances are measured, ultimately, against this goal. There is, however, no point in attempting to measure these utilities directly. What we are after is the maximally allowable values for α and β, and it seems much easier to make judgments about these probabilities than about the related utilities or expected utilities. An equally important question is, Who makes these judgments? In the end, I think, they must be made by people experienced in the field, though some judgments would be obvious to anyone with a little scientific common sense. In order to see more realistically just what kind of judgment is involved, let us look in greater detail at a typical statistical test of a simple causal hypothesis.

Imagine the beginning of an exploratory inquiry into the genetic effects of radiation on mammals. One might wonder whether a certain dose would affect the ratio of dominants to recessives in a given species. Suppose the crossing is such that the normal ratio is .5. The stage is then set for a UMPU test of the "null" hypothesis H: $p = .5$ versus the composite alternative, K: $p \neq .5$. Instead of a single number, β, the probability of Type II error, there is a whole function, the complement of the power

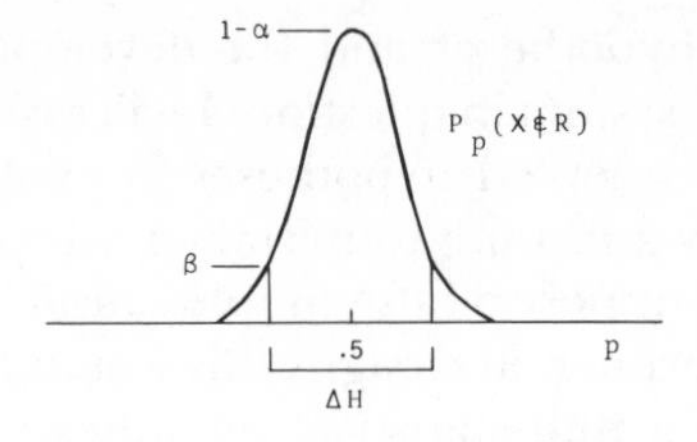

Figure 11

function, giving this probability for each simple component of K, as in figure 11. The question of which function is right for this particular test can be given a clear methodological meaning as follows: Consider values of p above and below .5 out to the point at which they are enough different from .5 to be worth further study. One way, though of course not the only way, of deciding on this point is to ask how great the difference would have to be for one to be willing to run only an α risk of not detecting it. Call ΔH the hypothesis that p is in the designated interval around $p = .5$. It is really ΔH, rather than H, that is accepted if H is not rejected. Here the two types of error are about as symmetrical as possible, given a composite alternative. The probability of accepting ΔH if false is no greater than the probability of accepting K if false (that is, rejecting H if true).[36] The sample size necessary for a satisfactory experiment is then determined once one sets the maximum allowable α.

Now, can scientists make sufficiently objective judgments on the values of α and ΔH for reasonable progress in an exploratory inquiry? I think they can, and even sometimes do. Unfortunately, practicing scientists still have not sufficiently assimilated the innovations of Neyman and Pearson, so there is still too much emphasis on significance levels and much too little consideration of power. The 5 percent significance level seems roughly appropriate for exploratory studies in biology—which was the context of Fisher's work. The fact that many psychologists complain about the tyranny of the 5 percent level is probably an indication that a higher level would be sufficient in much psychological research. The same is true, probably with even greater justification, in the social sciences. Only a close look at some representative lines of research in these fields could substantiate these claims. The best I can do here is offer an extreme hypothetical example.

Imagine a group of radically empiricist sociologists deciding to put their science on firm empirical foundations. They propose to edit a new journal whose editorial policy will be that all reported statistical tests must have significance levels of 1 percent or less and a probability of 99 percent of detecting 5 percent deviations from the null hypothesis. It is obvious that the project would be doomed from the start. Following this policy would lead to a fantastic waste of time and resources, gathering very large samples to detect small causal effects that can only confuse anyone looking for a simple but powerful theoretical model. There would be few Type I errors, to be sure, which means few assertions about nonexistent causal factors. But there is no reason to believe progress would be better than if α were raised to 5 percent or even 10 percent. And one does not need to be an expert in the field to make these judgments.

In recent years advocates of Bayesian information models for scientific research have claimed that "objectivist" statisticians are hypocritical, claiming complete objectivity while employing their own subjective judgment in setting parameters for statistical tests.[37] It might be thought that my appeals to judgments by scientists amount to admitting the soundness of these claims. This is not so. It is true that there is no such thing as a scientific epistemology that does not depend on informed judgment in some way. But it does not follow, as these theorists claim, that the role of such judgments is best or even adequately represented by subjective belief functions over admissible hypotheses. It is better represented, I think, in the above account of Neyman-Pearson testing models. Certainly the judgments one would want to make about the hypothetical radical sociologists do not concern the relative probabilities of any hypotheses. They concern the possibility of progress in a certain kind of inquiry.

To summarize, the rationality of Neyman-Pearson tests in exploratory inquiry consists in this: For any suspected causal factor in any type of system, the probability of the effect is changed either appreciably or only negligibly. In either case, the test is designed so that the probability of reaching the correct conclusion is sufficient for the purposes of exploratory inquiry.

5.3. Testing Causal Theories: Confirmatory Inquiry

I will now argue that the methodological conditions determining sufficient parameters for Neyman-Pearson statistical tests are different when the task is not to develop a theory but to test one already at hand. This will require some brief comments on what causal theories are, how they are tested, and why tests of theories involve tests of statistical hypotheses.

A theory provides a detailed description of a kind of system. It does so by utilizing a set of variables which represent various characteristics of the system. A set containing one value for each variable gives the state of the system at some time. In one form or another a theory must contain laws specifying which of the logically possible states are physically possible and also laws giving the dynamic evolution of the system in time. If the laws give a unique evolution from any initial state, the system is deterministic, like classical mechanics. But the laws may also give only probability distributions over later states, like Mendelian genetics.[38]

Just because theories attempt to describe the underlying causal structure of a system, the values of their variables are rarely open to direct inspection. These values can be ascertained only via more or less remote interactions. Here Mendelian genetics provides a good example. The theory is basically about genotypes and their transmission. The data, at least originally, consisted of observations on phenotypes. The two are mediated by a partial theory of development expressed in Mendel's laws of dominance. It is convenient to schematize these relationships as follows. First is the primary system, the object of the theory in question. Next is an intermediate system in interaction with the primary system. Finally there is a measuring system interacting with the intermediate system. Now the traditional hypothetico-deductive account of theory testing is correct to the extent that one step in a test of any theory is the deduction from the primary theory, together with the theories of the intermediate and measuring systems, and suitable initial conditions, of a hypothesis about observable states of the measuring system. What is not usually noted is that in most interesting cases, the "prediction" is really a statistical hypothesis concerning the output of the measuring system. This is obvious when the theory itself is probabilistic, as with Mendelian genetics. But it is also generally true even if all the theories involved are deterministic because measuring instruments behave like probabilistic systems at their limits of precision. The whole arrangement, with the example of Mendelian genetics, is diagrammed in figure 12. To make the example more specific, consider Mendel's backcross test, which is a cross between the first filial generation, all heterozygotes, and the recessive strain of the parental generation. The predicted ratio of dominants is .5. If Mendel is to be believed, that was a genuinely new prediction from the theory. He had not tried this combination before.

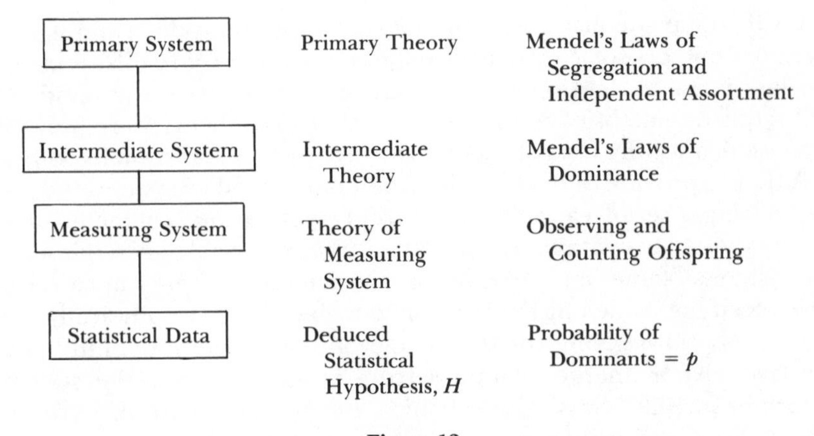

Figure 12

When a theory is being tested there are two tests proceeding simultaneously, the test of the theory and the test of the predicted statistical hypothesis. The two are obviously related in that whether the theory is accepted or rejected depends in a direct way on whether the statistical prediction is accepted or rejected. It is just this connection that provides an objective methodological basis for the choice of test parameters for the statistical test. The role of accepted statistical hypotheses in confirmatory inquiry is only to provide evidence for accepting or rejecting the theory. Thus, to accept a statistical hypothesis in confirmatory inquiry is to accept it as true for the purpose of testing a theory. Let us pursue this connection by looking at tests of theories in greater detail.

The general logical structure of a test of a theory is, I will argue, the same as that for a statistical hypothesis. With this perspective, attention is directed not primarily at the theory itself, but at the testing process. A good test of a theory, roughly speaking, will be a process that has a sufficiently low probability of rejecting the theory if it is true and a correspondingly high probability of rejecting it if false. The main process involved in testing a theory is that of selecting a statistical consequence of the theory which is itself tested against statistical data. The sample space for the test of the theory will then be the relative number of predicted statistical hypotheses that are accepted as true. Surprisingly, as few as five accepted independent predictions can provide a very good test of any theory, if one uses the most powerful decision rule for the theory test and corresponding parameters for the statistical tests.

To accept a theory is to take it as true for the purposes of explaining phenomena and guiding further investigation. It is not to take it as true for any particular practical decision. That requires further consideration of the evidence and the utilities of the practical context. With this restriction well understood, I think most scientists would regard theory testing error probabilities in the 1 percent range as quite acceptable, with 5 percent being the outer limit. These probabilities reflect the amount of effort it takes to construct any plausible theory and the value to theoretical progress of discovering that one's proposal is false, if it is. The problem, then, is to fix the error probabilities of the relevant statistical tests so as to achieve the desired range of error probabilities for the test of the theory.

The most powerful test of a theory is one that rejects the theory if any of its statistical consequences is false. Turning first to Type I error, if the primary theory is true and no false assumptions are introduced along the way, the predicted statistical hypothesis will be true. Thus, a true theory will be mistakenly rejected only if one of its statistical consequences is mistakenly rejected. Now if the theory were tested with only one prediction, the probability of a Type I error for the theory would be just α, the significance level of the statistical test of the predicted hypothesis. A 5 percent statistical test would give us a 5 percent chance of mistakenly rejecting a true theory. As will be clear presently, however, one prediction does not provide a very powerful test of any theory. Four or five are needed. But the probability of at least one mistaken rejection in five independent 5 percent tests is 23 percent—clearly outside the desirable 1 to 5 percent range. Five 1 percent statistical tests, on the other hand, yield only a 5 percent chance of mistakenly rejecting a true theory. For five .5 percent tests this probability is 2.5 percent, and for five .1 percent tests it is .5 percent. Thus, just by requiring a reasonably powerful test of the theory, we see that the significance levels of statistical tests in confirmatory inquiry can be no greater than 1 percent and should probably be less. This contrasts with exploratory inquiry where 5 percent tests, and sometimes even 10 percent tests, may be quite acceptable.

Four or five independent predictions are necessary to test a theory, because we want a 95 to 99 percent chance of rejecting the theory if it is false. As usual, the problem of mistaken acceptance is more complicated than the problem of mistaken rejection. Since a false theory may have true consequences, two things must happen if a false theory is to be correctly rejected as such. First, the consequence selected for testing must in fact be false. Second, the statistical test of the consequence must correctly reject it as false. Failure at either stage will lead to a Type II error in the test of the theory. Both stages pose special difficulties.

At the statistical level, the power against alternatives to the predicted statistical hypothesis must be great enough to detect clear deviations but not so great as to detect what might be only small systematic errors in the intermediate system, the measuring system, or in the initial conditions. For example, the predicted probability of dominants in the first backcross test is .5. But conditions in the development of the organisms could well make this probability .51 even if the genetic transmission of the trait in question is perfectly Mendelian. One would not want to reject the theory because of such a deviation. On the other hand, a probability of .6 for dominants is not likely unless the trait deviates from the simple Mendelian laws. For such deviations from prediction the power of the statistical test should be high, say 90 percent. Choosing the appropriate power function for testing statistical predictions thus requires reasonable knowledge of the intermediate and measuring systems involved in the test, but this is knowledge that is likely to be available by the time inquiry reaches the confirmatory stage.

The process of selecting consequences for testing is equally important if we are to have a high probability of rejecting false theories. Unfortunately this is a process which, in the very nature of the enterprise, we cannot know much about. If one selected for testing only consequences that had already been accepted in an earlier exploratory inquiry, that would be obviously unacceptable. Even if the theory is false, the probability of our now rejecting one of these "predictions" is small. However, even if we select only consequences not previously investigated, like Mendel's backcross, we have no way of knowing the probability that we have selected a false consequence if the theory is false. The best we can do is exploit all our scientific intuition and insight in an attempt to find consequences that might be false. We still will have no way of knowing how successful we are, but this will not matter too much as long as we are not terribly unsuccessful. For example, suppose our propensity for selecting clearly false consequences from a false theory is only 50 percent. And suppose the power of our statistical tests against such clearly false deviations is 90 percent. Then the probability of our rejecting a false theory on the basis of one predicted statistical hypothesis is only 45 percent. But with five such predictions, the probability of rejecting one prediction, and thus rejecting the theory, rises to 95 percent. This is adequate. On the other hand, if our propensity for picking out false consequences from a false theory is only 25 percent, then our probability for rejecting the theory with five predictors cannot be more than 76 percent, no matter how powerful our statistical tests against clearly deviant alternatives. Thus, it is important that we try hard to find false predictions.[39]

Proponents of the testing paradigm have long held that successful predictions of new phenomena provide much better evidence for a theory than mere agreement with previously known phenomena.[40] Proponents of the information paradigm have never been able fully to appreciate this doctrine, since the same amount of information is involved whether it comes sooner or later. This account of theory testing provides a clear justification for this doctrine. A good test of a theory is a process that has low probabilities for leading us to reject a true theory or to accept a false one. Now consider the process of constructing a theory subject to the constraint that any seriously proposed theory agree with a given set of accepted statistical hypotheses. The best one could say for the error probabilities of this process is that they are unknown. On general empirical grounds it would seem that they are in fact quite high. In particular, to devise a process with a good chance of leading to the rejection of a false theory, one needs a selection process with a reasonable probability for selecting a false consequence. Selecting hypotheses already successfully tested is obviously not such a process. Selecting untested consequences which experienced investigators suspect may be wrong is surely better and, while not provably sufficient, clearly the best we can do.

In summary, the methodology of theory testing provides grounds for fixing the parameters of statistical tests incorporated in the theory-testing context. The decision principle behind this methodology is satisficing. The statistical consequences of the theory selected for testing may be very close to correct or not. In either case, the parameters of the statistical tests are adequate for the task of testing the theory at issue. Moreover, roughly the same methodological principle also applies at the level of theory testing. The theory in question may be approximately true or not. In either case the probability that we reach the right conclusion is adequate for the purposes of theoretical science—or at least as good as we can make it.

6. CONCLUSION: SCIENTIFIC KNOWLEDGE AND PRACTICAL DECISION-MAKING

An essential ingredient of my attempt to provide a methodology for the application of the basic Neyman-Pearson logic was restricting the application to specific types of scientific contexts. It is the scientific context that provides certain kinds of utilities that are relatively stable over many applications of Neyman-Pearson tests and that can be indirectly estimated by experienced investigators.[41] Scientific knowledge thus consists of simple causal hypotheses and theories that are accepted as true for the purposes of a particular kind of scientific inquiry. It is possible, therefore, that while a hypothesis is acceptable as true in a scientific context, the error probabilities are too large for one to risk acting on that assumption in a particular practical context. What, then, is the relation between scientific knowledge, so conceived, and practical decision-making?

The answer depends on one's account of rational decision-making. My own view is that different principles—for example, Bayes' rule, cost-benefit, satisficing—will be appropriate in different contexts, depending in large measure on the nature of the available information. But then one needs principles, which currently are unavailable, for judging which method to apply. For present purposes, I will focus on Bayes' rule.

One of the major attractions of Bayesian information models of statistical inference is that they provide a clear account of the relation between scientific knowledge and practical action. All the scientific information is contained in the probabilities of the hypotheses. To make a decision one merely adds one's utilities and applies Bayes' rule. Now an "objectivist" can use Bayes' rule as well, but with the restriction that the probabilities employed be physical propensities, or rather, estimates of the values of physical propensities. This restriction places strong limitations on the use of Bayes' rule and indicates to some extent how the scientific and practical contexts are related.

In addition to models for *testing* statistical hypotheses, Neyman and others have developed parallel models for *estimating* statistical parameters. I have emphasized testing models in this essay because typical scientific contexts make the use of tests common and appropriate. We want to know whether a particular factor has a nonnegligible effect on some other or whether some predicted statistical hypothesis is approximately true. This is not to say that there are no scientific reasons for estimating parameters as well. For practical contexts, however, this emphasis is reversed and exaggerated. There is little use for tests as such. What we want are estimates of statistical parameters, because decisions depend on what their values are, not on whether they are close to some null or predicted value.

There is no need here to go into the details of Neyman's account of parameter estimation.[42] The most important general point is that for Neyman, estimation, like testing, is a process, so that what makes a good estimate is an estimation process with good characteristics. There is no direct evaluative relation between data and the estimate. To illustrate, suppose we merely want to estimate the probability, p, of dominants in some particular cross. Using the basic formal apparatus of hypothesis testing, Neyman has shown how to construct an interval of values of p as a function of the observed frequency of dominants in some sample. This "confidence interval" is constructed in such a way that it has an assignable probability, α (the "confidence level"), of including the true value of p, no matter what this value may be. Note in particular that this does not mean that p has probability α of being in the interval. Such statements presuppose a probability distribution over values of p—a distribution that Neyman does not assume.

Now imagine a decision problem, for example, that of a plant breeder, which depends on the ratio of dominants. If the issue arises in the practical context, there is no particular reason to expect that any data on this cross have ever been gathered. That depends on this particular cross having earlier been of scientific interest in some kind of scientific inquiry. There are no special connections between what will be of scientific interest and what will be of practical interest.

Suppose, however, that some data on the cross had been gathered in order to test the hypothesis that $p = .5$. The practical decision maker can ignore the scientist's conclusion and simply use the data themselves to construct the shortest 95 percent confidence interval for values of p. Suppose this interval is $.45 \leq p \leq .65$. Unfortunately it might turn out that Bayes' rule with $p=.45$ favors not breeding that strain, while with $p=.65$ it favors breeding. In this case it is not obvious what to do. On the other hand, if more data had been gathered, one might have been able to construct a 99 percent confidence interval for p with $.55 \leq p \leq .60$. And it might turn out that Bayes' rule favors breeding for any p in the whole interval. In this case the decision to breed the strain seems obviously justified. But there is no reason why the amount of data that is appropriate to a scientific test (or a scientific parameter estimate) should be sufficient to lead to a clear-cut decision in some particular decision problem. The two activities are just not that closely related.

The moral to be drawn, as governments and large industrial corporations have known all along, is that good, objective decision-making requires a large investment in "research and development" explicitly directed at specific practical problems. Those of us without access to such resources have less desirable options. We can hope that there already exist data yielding estimates precise enough to produce clear choices by maximizing objectively expected utilities. This might come about as a result either of scientific or of practically oriented research. Or, lacking the necessary data, we can fall back on subjective estimates of the relevant propensities. This is not to adopt a subjective information model of decision-making, since the distinction between objective and subjective estimates is retained. And what is estimated are physical propensities. Using subjective estimates when one must reach some decision in the face of insufficient data is simply relying on the fact that people with some experience, however vague, can usually do better using such estimates then merely choosing randomly.

To pursue these issues any further would take us too far into an area which, in my view, is really quite distinct from the subject of this essay, though of course it is related. I have said enough, I hope, to show that my analysis of scientific inference allows some connections between scientific knowledge and practical action, though less direct than usually supposed. And if it leaves practical decision-making a rather messy affair, this should not be surprising. The practical side of life is a messy affair.

NOTES

The author's work on the foundations of statistics has been supported in part by the National Science Foundation.

1. The influence of Einstein and modern physics on Popper's thought is perhaps most evident in the sometimes very personal essays reprinted in *Conjectures and Refutations* (New York: Basic Books, 1962).

2. The confirmation of laws, however, has been a central topic among confirmation theorists in recent years. The most extensive treatment of the problem, by a student of Hintikka's, is Juhani Pietarinen's *Lawlikeness, Analogy, and Inductive Logic* (Amsterdam: North-Holland, 1972). For further references on standard and recent work in confirmation theory see Pietarinen's bibliography and also Richard Swinburne, *An Introduction to Confirmation Theory* (London: Methuen, 1973). My own views on the prospects for any confirmation theory based on logical probabilities are expressed in my review of volume 1

of *Studies in Inductive Logic and Probability,* ed. Rudolf Carnap and R. C. Jeffrey (Berkeley: University of California Press, 1971), which appeared in *Synthese,* 23 (1975): 187–99.

3. I have in mind chapters 5–8 of R. B. Braithwaite's *Scientific Explanation* (Cambridge: Cambridge University Press, 1953) and C. West Churchman's *Theory of Experimental Inference* (New York: Macmillan, 1948). Churchman's book was much less influential than it should have been, perhaps because Churchman was outside the logical empiricist tradition that came to dominate philosophy of science in the United States after World War II. And Churchman himself moved further outside the mainstream of philosophy into decision theory and operations research.

4. See Ian Hacking, *Logic of Statistical Inference* (Cambridge: Cambridge University Press, 1965) and A. W. F. Edwards, *Likelihood* (Cambridge: Cambridge University Press, 1972). The fact that I do not discuss likelihood models in this chapter reflects my judgment that likelihood, by itself, cannot provide an adequate basis for a comprehensive account of statistical inference.

5. References to Savage will appear frequently in this chapter. Ramsey's "Truth and Probability" and de Finetti's well-known "Foresight: Its Logical Laws, Its Subjective Sources" are reprinted in *Studies in Subjective Probability,* ed. H. E. Kyburg, Jr., and H. E. Smokler (New York: Wiley, 1964). The definitive source for de Finetti's views is now his two-volume *Theory of Probability* (New York: Wiley, 1974–75). More readable are some of the essays in the collection *Probability, Induction and Statistics* (New York: Wiley, 1972).

6. Fisher, of course, also published numerous papers in mathematical statistics, a good selection of which is reprinted in R. A. Fisher, *Contributions to Mathematical Statistics* (New York: Wiley, 1950).

7. Any good text book of .genetics will contain a discussion of Mendelian theory, for example, chapters 6 and 7 of M. W. Strictberger, *Genetics* (New York: Macmillan, 1968).

8. The quotation is from the 14th edition of *Statistical Methods,* p. 80. This edition appeared in 1970 and contains Fisher's last revisions. The larger context of the quotation clearly shows the influence of Fisher's thinking around 1957, when he wrote *Statistical Inference and Scientific Method,* some thirty years after the first publication of *Statistical Methods.* The corresponding passage in the 3rd edition (1930) is more ambiguous. It does not contain the phrase "strength of evidence against the hypothesis." Sorting out Fisher's views on statistical inference at various periods of his career is clearly a very difficult task.

9. For example, Ian Hacking, in *Logic of Statistical Inference,* and more recently Stephen Spielman, ' The Logic of Tests of Significance," *Philosophy of Science,* 41 (1974):211–26.

10. R. A. Fisher, *The Design of Experiments,* 4th ed. (New York: Hafner, 1947), p. 13.

11. Kenneth Mather, *The Measurement of Linkage in Heredity,* 2d ed. (London: Methuen, 1951), p. 7.

12. Fisher, *Design of Experiments,* p. 16.

13. The crucial papers, published during the decade between 1928 and 1938, are now reprinted in J. Neyman and E. S. Pearson, *Joint Statistical Papers* (Berkeley: University of California Press, 1967). For a personal account of these years see Egon Pearson's "The Neyman-Pearson Story: 1926–34," in *Research Papers in Statistics,* ed. F. N. David (New York: Wiley, 1966). Elements of the Neyman-Pearson account of statistical hypothesis testing are treated in almost any textbook of pure or applied statistics. Conceptually, one of the best treatments is Neyman's own text which, unfortunately, is now out of print. See J. Neyman, *First Course in Probability and Statistics* (New York: Henry Holt, 1950). The standard, modern, advanced text is E. Lehman, *Testing Statistical Hypotheses* (New York: Wiley, 1959). For a much more elementary presentation, see the latter chapters of J. L. Hodges and E. L. Lehman, *Basic Concepts of Probability and Statistics* (San Francisco: Holden and Day, 1964).

14. See the very last section of Neyman's *First Course.*

15. Neyman's clearest statement of these views is in his "'Inductive Behavior' as a Basic Concept of Philosophy of Science," *Review of the International Statistical Institute*, 25 (1957):7–22.

16. Of Wald's writings, the most accessible intellectually is unfortunately not the most accessible physically. This is Abraham Wald, *On the Principles of Statistical Inference*, Notre Dame Mathematical Lectures, 1 (South Bend: University of Notre Dame, 1942). The most comprehensive and systematic of his works, *Statistical Decision Functions* (New York: Wiley, 1950), is mathematically quite advanced. The best elementary presentation that retains the spirit of Wald's approach is Herman Chernoff and Lincoln E. Moses, *Elementary Decision Theory* (New York: Wiley, 1959).

17. The next two sections draw heavily on the treatment given in R. D. Luce and Howard Raiffa, *Games and Decisions* (New York: Wiley, 1957).

18. See, for example, J. W. Milnor, "Games Against Nature," in *Decision Processes*, ed. R. M. Thrall, C. H. Coombs, and R. L. Davis (New York: Wiley, 1954).

19. For a discussion of various axioms, their consequences and logical relations, see chapter 13 of Luce and Raiffa's *Games and Decisions*.

20. The form of the example seems to have originated with D. R. Cox. See his "Some Problems Connected with Statistical Inference," *Annals of Mathematical Statistics*, 29 (1958):357–72. This same example is presented as a major criticism of "Classical Statistical Theory" by Henry E. Kyburg, Jr., in his recent major work, *The Logical Foundations of Statistical Inference* (Dordrecht, Holland: Reidel, 1974). I have also discussed the example in section 2 of "Empirical Probability, Objective Statistical Methods, and Scientific Inquiry," in *Foundations of Probability Theory, Statistical Inference and Statistical Theories of Science*, vol. 2, *Foundations and Philosophy of Statistical Inference*, ed. W. L. Harper and C. A. Hooker (Dordrecht, Holland: Reidel, 1976), pp. 63–101.

21. Here I have in mind Wesley C. Salmon's work as presented in *The Foundations of Scientific Inference* (Pittsburgh: University of Pittsburgh Press, 1967) and *Statistical Explanation and Statistical Relevance* (Pittsburgh: University of Pittsburgh Press, 1971).

22. Recent advocates of some kind of propensity interpretation, though not necessarily by that name, include: R. B. Braithwaitę, Mario Bunge, James H. Fetzer, Ronald Giere, Donald Gilles, Ian Hacking, Isaac Levi, D. H. Mellor, Karl R. Popper, and Tom Settle. In addition one should mention C. S. Peirce, whom many regard as the original propensitist, and the statistician Harald Cramer.

23. The reference here is to the line of work begun by Robert Stalnaker and Richmond Thomason, for example, "A Semantic Analysis of Conditional Logic," *Theoria*, 36 (1970): 23–42; and David Lewis in *Counterfactuals* (Oxford: Blackwell, 1973).

24. Henry E. Kyburg, Jr., "Propensities and Probabilities," *British Journal for the Philosophy of Science*, 24 (1974):358–75.

25. See L. Sklar, "Is Probability a Dispositional Property?", *Journal of Philosophy*, 67 (1970):355–66; and "Unfair to Frequencies," *Journal of Philosophy*, 70 (1973):41–52.

26. I have argued for the usefulness of a conception of absolute propensities in "Objective Single Case Probabilities and the Foundations of Statistics," in *Logic, Methodology and Philosophy of Science*, vol. 4, ed. P. Suppes et al. (Amsterdam: North Holland, 1973), pp. 467–83; and in "A Laplacean Formal Semantics for Single-Case Propensities," *Journal of Philosophical Logic*, 5 (1976).

27. See Hans Reichenbach, *The Theory of Probability* (Berkeley: University of California Press, 1949); and Wesley C. Salmon, *Foundations of Scientific Inference* and *Statistical Explanation and Statistical Relevance*. A surprising amount of what both Reichenbach and Salmon have had to say about these problems is clearly foreshadowed in John Venn's *Logic of Chance* (London, 1866).

28. These points are argued in the two papers cited in note 26. I have also made a lengthy attempt to construct a nonfoundationist justification of inductive inference based

on Neyman-Pearson testing models of statistical inference. See "The Epistemological Roots of Scientific Knowledge," in *Induction, Probability and Confirmation Theory,* Minnesota Studies in the Philosophy of Science, vol. 6, ed. Grover Maxwell and Robert M. Anderson (Minneapolis: University of Minnesota Press, 1975). The present essay may be regarded as an attempt to resolve the problem of characterizing inductive logic described in the introduction to this latter paper.

29. The original sources of the argument are pp. 170–75 of L. J. Savage, "Bayesian Statistics," in *Recent Developments in Decision and Information Processes,* ed. R. E. Machol and P. Gray (New York: Macmillan, 1962); and pp. 13–15 of Dennis Lindley, *Bayesian Statistics: A Review* (Philadelphia: Society for Industrial and Applied Mathematics, 1971). To the best of my knowledge, the expression "The Lindley-Savage Argument" originated in conversations between Allan Birnbaum and myself during the academic year 1971–1972. Birnbaum has subsequently presented an analysis of the argument in an as yet unpublished paper, "The Neyman-Pearson Theory as Decision Theory, and as Inference Theory; with a Criticism of the Lindley-Savage Argument for Bayesian Theory." The following presentation of the argument draws heavily on Birnbaum's analysis. Our responses to the argument, however, are significantly different.

30. The above is an adaptation and expansion of the argument Lindley gives in *Bayesian Statistics: A Review,* pp. 13–15.

31. Savage cites a paper by E. L. Lehman, "Significance Level and Power," *Annals of Mathematical Statistics,* 29 (1958):1167–76.

32. This is in fact what Lindley does say in *Bayesian Statistics: A Review,* p. 14. Savage, however, says: "Insofar as this demonstration carries weight, it makes you a Bayesian whether you thought you wanted to be or not." See "Bayesian Statistics," p. 175.

33. The classic discussion of the distinction is in section 44 of Rudolf Carnap's *Logical Foundations of Probability* (Chicago: University of Chicago Press, 1950). The usefulness of the distinction, however, is not restricted to the application of theories of logical probability. Indeed, one of the foremost current defenders of the distinction is Wesley Salmon, also a prime opponent of the use of logical probabilities. See his *Foundations of Scientific Inference.*

34. The best discussion of satisficing I know is by Herbert Simon, who originated or at least first popularized the notion. See pp. 196–206 and chaps. 14 and 15 of *Models of Man, Social and Rational* (New York: Wiley, 1957). There is more literature on the subject in the areas of economic and business decision-making. The only roughly philosophical discussion I know is Alex Michalos, "Rationality Between the Maximizers and the Satisficers," *Policy Sciences,* 4 (1973):229–44.

35. This sort of problem was first emphasized, in the philosophical literature at least, by Richard Jeffrey, in "Valuation and the Acceptance of Scientific Hypotheses," *Philosophy of Science,* 23 (1956):237–46, Jeffrey's argument has strongly influenced other students of scientific inference, for example, Isaac Levi in *Gambling with Truth* (New York: Knopf, 1967). I have discussed the problem in somewhat more detail at the beginning of section 3 of "Empirical Probability, Objective Statistical Methods and Scientific Inquiry."

36. In introducing the spread-out null hypothesis, ΔH, I am not adding anything to the Neyman-Pearson formalism. It is just a reinterpretation of the role of the power function. It also provides a vindication for the intuition that no amount of evidence could justify the conclusion that some suspected causal factor has literally no effect whatsoever. I have discussed the acceptance of ΔH in connection with Popper's doctrine that inductive acceptance of any hypothesis is impossible in "Popper and the Non-Bayesian Tradition," *Synthese,* 23 (1975):119–32.

37. For example, Savage, in "Bayesian Statistics."

38. The account of theories I have in mind here is not the traditional one but something like the "semantic" approach advocated by Bas van Fraassen in "A Formal Approach to the

Philosophy of Science," in *Paradigms and Paradoxes,* ed. R. G. Colodny (Pittsburgh: University of Pittsburgh Press, 1972).

39. Thus, we have an "inductivist" justification for Popper's doctrine that testing a theory requires that we try to falsify it. I have discussed this point in "Popper and the Non-Bayesian Tradition." A number of authors have offered Bayesian justifications of this same doctrine. See Wesley Salmon's *Foundations of Scientific Inference* and Richard C. Jeffrey, "Probability and Falsification," *Synthese,* 23 (1975):95–117. My justification is both inductivist and non-Bayesian.

40. The most recent discussion of this doctrine, with favorable quotations extending back to the beginning of the seventeenth century, is Alan E. Musgrave's "Logical Versus Historical Theories of Confirmation," *The British Journal for the Philosophy of Science,* 25 (1974):1–23. Musgrave is quite mistaken, however, in focusing on differences between logical and historical accounts of the confirmation of theories. The justification given earlier is neither nonlogical nor essentially historical (except that it does not use logical probabilities and does consider the process of testing). The point is simply that predictions of as yet uninvestigated hypotheses generally (though not necessarily) provide more powerful tests (in the Neyman-Pearson sense) of theories.

41. The idea that there might be scientific or epistemic utilities was discussed by C. G. Hempel in "Deductive-Nomological vs. Statistical Explanation," in *Scientific Explanation, Space and Time,* Minnesota Studies in the Philosophy of Science, vol. 3, ed. H. Feigl and G. Maxwell (Minneapolis: University of Minnesota Press, 1962). Epistemic utilities also play a central role in Isaac Levi's views as expressed in *Gambling with Truth.* All the attempts I know to use epistemic utilities, including these, assume either logical or subjective probabilities over hypotheses and rely on Bayes' rule. And the utilities invoked—for example, content—are, like the probabilities, assigned to particular hypotheses. In keeping with the spirit of the testing paradigm of scientific inference, the utilities in my analysis attach to classes of tests in a whole process of inquiry, and not directly to individual hypotheses.

42. The theory of estimation by confidence intervals is discussed in most texts of mathematical statistics. For a really good discussion see volume 2 of M. G. Kendall and A. Stuart, *The Advanced Theory of Statistics* (London: Griffin, 1958). The original source is Neyman's "Outline of a Theory of Statistical Estimation Based on the Classical Theory of Probability," *Philosophical Transactions of the Royal Society of London,* A 236 (1937):333–80.

ROBERT E. BUTTS
The University of Western Ontario

Consilience of Inductions and the Problem of Conceptual Change in Science

> If, in our induction, every individual case has actually been present to our minds, we are sure that it will find itself duly *represented* in our final conclusion: but this is impossible for such cases as were unknown to us and hardly ever happens even with all the known cases; for such is the tendency of the human mind to speculation, that on the least idea of an analogy between a few phenomena, it leaps forward, as it were, to a cause or law, to the temporary neglect of all the rest; so that, in fact, almost all our principal inductions must be regarded as a series of ascents and descents, and of conclusions from a few cases, verified by trial on many. . . . The surest and best characteristic of a well-founded and extensive induction, however, is when verifications of it spring up, as it were, spontaneously, into notice, from quarters where they might least be expected, or even among instances of that very kind which were at first considered hostile to them. Evidence of this kind is irresistible, and compels assent with a weight which scarcely any other possesses.
>
> —Sir John F. W. Herschel
> *A Preliminary Discourse on the Study of Natural Philosophy,* 1830

I. In recent papers by L. Laudan, M. B. Hesse, and others, and in my "Whewell's Logic of Induction," interesting and perhaps important questions have been raised concerning the concept of consilience of inductions.[1] The name of the concept, if not the concept itself, was invented by William Whewell, who made quite extravagant claims for cases of historical consilience (for example, the Newtonian synthesis), holding that consilience of hypotheses constitutes a test of truth, and that whenever consilience occurs the laws in the resulting theories are necessary truths. Whewell's own idiosyncratic metaphysics and epistemology attempted justifications of these claims, but I do not find evidence that any recent writers are disposed to become full-bodied disciples of Whewell. Apart from a certain inevitable aesthetic appreciation of the intricacies of his full metaphysics of science, there is little in Whewell's complete system that has philosophical appeal for today's philosophers of science. Whewell's writings abound in large and unrecoverable logical slips (of some of which he was enormously proud),[2] and often in reading him one awakens from the sweet sleep induced by his Victorian prose and desperately looks for an argument, only to find a kind of suggestive philosophical poem instead.

Realizing these peculiarities of Whewell's own philosophy, none of the recent writers has attempted to revive Whewell's inductive logic as a viable alternative to any of today's logics. Laudan's task was to show that Whewell's concept of consilience entails that the most important consiliences occur where a certain maximally acceptable level of corroboration of a theory has been reached, thus attempting to save one crucial empirical factor in a philosophy of science that is otherwise seventeenth-century rationalistic through and through. Hesse's interest has been in seeing whether she can reconstruct something like Whewell's consilience in the context of normal probabilistic confirmation theory; and she has concluded that confirmation theory alone will not capture the Whewellian notion that consiliences increase confirmation of laws. In my earlier paper, the analysis of consilience that I give seems to me to accord pretty well with Hesse's conclusions about consilience, though I also agree with Laudan that there is something important about the concept that is worth retaining and talking about. That paper argues that Whewell's inductive logic actually involves two logics, one the standard hypothetico-deductive method (of which Whewell's exposition is masterful), the other a merely suggested methodology that would count consiliences of inductions as marks of the acceptability of laws or theories. In the second methodology, consilience counts along with simplicity as a measure of the acceptability of a theory.

Of course, one runs into problems on my line of analyzing Whewell's inductive logic. Whewell clearly accepted the hypothetico-deductive method—for him induction is the inverse of deduction—and insisted to the end that all proper scientific systems be cast in deductive form. Given this fact, coupled with our present understanding of the defects in the hypothetico-deductive methodology,[3] we should conclude that Whewell's references to consiliences as tests of truth should be read as suggesting that consilience, like simplicity, is an *extraevidential* test of theories. However, in an important recent paper, Wesley Salmon has suggested that we move too quickly to the conclusion that what he calls plausibility considerations (I am going to argue below that consilience is a major plausibility consideration) must be taken as extraevidential, just because of failings in the hypothetico-deductive methodology, and because of failures to fit concepts like consilience into standard probabilistic confirmation theory. His alternative suggestion is that we change our concept of *evidence.*[4]

Salmon's point, briefly put, is that the specification of prior probabilities needed to get Bayesian confirmation theory to work is bound up with the *plausibility* of hypotheses. I will not argue here the question of whether Salmon's move on this point is sound, nor will I discuss the point that plausibility considerations may themselves introduce irreducible extraevidential factors. Salmon's interesting move—the move that I think may help us with a more accurate reconstruction of Whewell's employment of the concept of consilience—is to argue that the prior success of scientific theories, a success of historical standing, can itself be taken as part of the "evidence" for new discoveries or new inductive (Bayesian) moves, in that this success renders certain hypotheses more antecedently probable, that is, more plausible.[5] Salmon's point is all the more interesting in that, though he fully accepts the distinction between the context of discovery and the context of justification, he is suggesting a way in which the data of the history of science can come to play an important role in probabilistic induction. All of this fits neatly with Whewell's claim that his inductive logic was a "Discoverer's logic" and with his insistence that the history of science is in some important way the justification of science, a point which, by the way, is admirably expressed in section III of Laudan's paper.

One additional important point has emerged in the recent discussions of consilience. Mary Hesse has made the point by arguing that in order for consilience to have any logical integrity, deductive entailment of laws and data-sentences is neither sufficient nor necessary, but in addition we must be able to note relevant analogies between consilient laws. My way of making the point is to suggest that in every case of consilience, in order to get deductive connections between parts of a new theory and an older one, certain fundamental changes in the semantics of the older theory must be made. Thus, to understand how consilience operates in the history of science is to understand something about how scientific concepts change.

In what follows, then, I propose to try to do three things. First, to show where I stand on the issue that divides Laudan and Hesse, and in so doing argue that we must finally take Whewell's reference to consilience as a test of truth *in any hard epistemological sense as a mere façon de parler*. In other words, I will try to show that the fact that hypotheses become consilient adds nothing new or important to the confirmation or corroboration of these hypotheses, or, what amounts to the same thing, that the concept of consilience adds nothing to standard confirmation theory that we did not already know. Second, by expanding upon the discussion of consilience in my earlier paper, I will sketch the role that semantic or conceptual change plays in cases of historical consilience. Third, I will try to move from these two points to outline an interpretation of consilience (and one that I think fits best Whewell's requirements for the concept) by adopting Salmon's suggestion that historical success of theories be taken as evidential in the sense of making some hypotheses more antecedently probable than others, and by showing that it was something like this that Whewell was claiming for consilient theories when he took them to be incontestably true.

II. Whewell gives various characterizations of the concept of consilience, but one of his briefer statements will do for our purposes:

> The Consilience of Inductions takes place when an Induction, obtained from one class of facts, coincides with an induction, obtained from another class. This Consilience is a test of the truth of the Theory in which it occurs.[6]

For Whewell, the most prominent cases of such consilience are great scientific syntheses like the one achieved by Newton's inverse-square law. The thing to be remarked is the great variety of *apparently* different kinds of data that such a law comes to explain under one set of concepts. It should also be noted that consilience is not a property of all inductions. Consilience is a property of some, and only some, logically well-articulated theories that possess large measures of deductive content or that contain predicates that are expressively very rich. Consilience is thought to be a property of those systems having the following characteristics: (1) the theories must be so general that they have almost reached the point of unity; this for Whewell is equivalent to saying that the theories must be simple; (2) the theories must provide the best explanation of the large range of objects involved; and (3) the theories must have achieved that historical situation where further testing of the laws is seen to be irrelevant to acceptance of the theories; the theories must have attained the position where negative results will be taken as calls for refinement of the systems, rather than as disconfirmations.[7] In my earlier paper I provided a formal model for capturing consilience in what I take to be Whewell's sense. I reproduce it here to initiate discussion of the important problems. In section IV of this essay I will suggest that the model, though faithful to one Whewellian sense of consilience, will not do the required job.

Given two evidence classes E_1 and E_2, and two laws L_1 and L_2, at time t_1, L_1 explains E_1, and L_2 explains E_2, and there are no inductive reasons for supposing that L_1 and L_2 are connected; put differently, at t_1 it is *thought that* E_1 and E_2 are disjoint. (In this model, "explanation" means classical hypothetico-deductive explanation.)

Now at some later time t_2 a theory T is introduced. L_1 and L_2 become consilient at t_2 with respect to T when they both become logically derivable from T, that is, when T hypothetico-deductively explains both L_1 and L_2, and, derivatively, explains E_1 and E_2. At t_2, then, E_1 and E_2 are no longer thought to be disjoint. (The same model holds, of course, if L_1 and L_2 are laws offered in explanation of *other laws.*) We cannot allow the scientist an easy victory here; we cannot allow that T be the conjunction of L_1 and L_2, otherwise we would get as many consiliences as there are dogs in southern Portugal.

All writers agree that Whewell was on to something important at this point. All classical fanciers of the hypothetico-deductive theory of science had talked about science as an ever enlarging class of generalizations, which class itself would one day be caught at last by the biggest and best of all generalizations. But not many of them had attended properly to the crucial epistemological question of how we go about choosing among all of the logically possible biggest and best generalizations. Whewell was aware that induction at the two lower levels of confirmation (simple colligation of sets of events, relations between objects, or properties of all objects of a certain kind, and successful prediction of the same kinds of events, relations, or properties) was not sufficient to render a scientific theory fully acceptable. Indeed, I think I have shown that for Whewell the two lower kinds of induction are self-guaranteeing as to truth.[8] We want scientific theories that extend beyond the sets of data they were originally introduced to explain. Whewell, in introducing the concept of consilience, is noting a very special feature that is a property of all maximum-information-seeking scientific syntheses. The key feature of such theories is that they achieve *reductions* of a certain very special kind; in truly explaining kinds of data that were thought to be quite distinct and independent, these theories in fact reduce the apparently disparate kinds of data to one kind. This is Hesse's point when she argues that deductive connections between theories and laws are not sufficient to capture the idea of consilience; to get consilience we must be able to note significant analogies between the kinds of things talked about by the laws.[9] On the same point, Laudan correctly remarks that "the real strength of such an [consiliative] hypothesis is usually that *it shows that events previously thought to be of different kinds are, as a matter of fact, the 'same' kind of event.*"[10] In part III of this essay I will discuss this crucial point as it bears on the question of conceptual change in science. Just now, I want to show how this point entails that consilience does not add anything to our standard concepts of how theories get to be confirmed.

As I read their papers, the major difference between Hesse and

Laudan is the question of whether successful consilience increases our confidence in the *theories* with respect to which laws become consilient, or increases our confidence in the *laws* that become consilient with respect to a certain theory. Both authors are working in the context of trying to reconstruct the concept of consilience within normal probabilistic confirmation theory. Laudan also questions Hesse's "use" of the term consilience, claiming that her employment of the term is not Whewell's. I think this issue is quite incidental to the logical and epistemological issue that apparently divides them. As Hesse formulates the difference, there are two problems: (1) What Laudan takes to be Whewell's problem: T is suggested to explain L_1 and is supported by L_1; how much is our confidence in T increased if it also predicts L_2, and L_2 is *subsequently* directly confirmed? (2) Hesse's problem: T is suggested to explain L_1 and is supported by L_1; T also entails L_3; how much is our confidence that L_3 *will* turn out to be a correct prediction increased by the fact that T entails it?[11]

Given the formal model of consilience that I just proposed, I think that both problems (1) and (2) have an easy solution, and that the same solution suffices for both. Paradoxically, application of the model of consilience to these problems also argues for the eliminability of the concept of consilience as a special kind of confirmation or corroboration measure.

Take the Laudan formulation of Whewell's problem first. If T predicts L_2 at t_1, and we have no evidence that L_2 is supported by some E_2 at t_1, then we seem to have the standard problem of discovering whether or not L_2, as a prediction of T, is true, and thus confirms T. Obviously if L_2 turns out, on evidence, to be highly probable, then our confidence in T will increase to some extent, *but this increase in confidence will be quite independent of the evidence for T that comes from* L_1, *if* L_1 *and* L_2 *are logically independent hypotheses.* On the other hand, if L_1 and L_2 are not logically independent, as they may not be if both turn out to be true entailments of T, then the domains of evidence over which they range will, as Hesse points out, at least partly overlap; and the question of whether this overlap obtains cannot be decided in any way that I can see within the confines of probabilistic confirmation theory. Thus, the solution to Laudan's problem can be gotten independently of questions of consilience, and if consilience does take place, all the consilience shows is the way in which the domains of evidence of the two laws overlap. In other words, if L_1 and L_2 turn out to be confirmed entailments of T, at least some of the evidence for both will be held in common, and T will reduce at least parts of two evidence classes once thought to be distinct to a new overlapping domain of events, relations, or properties *of the same kind.* The reduction, however—and this is the crucial point—will not be achieved by ordinary confirmation theory considerations. The trouble, I think, lies in formulating Whewell's problem in deductivist terms, when in fact what is required is a view of theories that explicates the relationship between those theories and the laws that they involve in altogether different terms.

Thus, Laudan's formulation seems to me to show that consilience is irrelevant as an addition to ordinary confirmation theory. The major difference between Hesse and Laudan is that she publicly acknowledges this point. On her own analysis of what she calls her problem, she shows conclusively that our confidence that L_3 will turn out to be a true prediction of T is entirely bound up with the question of the relationship between that law and L_1. For her, it is not a question of whether or not the relevant laws are subsumed by T, but a question of the prior probability relation between L_1 and L_3.[12] And that relation, on her account, has to do with a different conception of the role of T in cases of consilience. T need not, indeed in most historical cases does not, entail the laws; in fact, even if T does entail the laws, this is incidental. What counts is that T function as the vehicle for relating L_1 and L_3 by pointing out relevant analogies, that is, showing the respects in which the evidence domains of L_1 and L_3 are the same, or analogous. But if T gives enough confidence that the two laws cover analogous domains, then of course our confidence in L_3 is increased. *At the same time,* our confidence in T is increased; that is, if T shows that an already confirmed L_1 is talking about evidence classes of the same kind as the as yet untested L_3, the probabilities of both T and L_3 seem increased. Thus, Hesse concludes that under the condition of construing a theory as playing the role of pointing out the relevant analogies between L_1 and L_3, problems (1) and (2) are solved together.

There is, I think, an important connection between Hesse's treatment of consilience and Salmon's suggestion that we take historical success of scientific theories as expressing the initial plausibility of scientific hypotheses. I will return to this point. For the present, I hope to have shown that where consilience has taken place, we are left with exactly the same probabilistic confirmation problems that we have always had with respect to determining extent of confidence in scientific hypotheses; and where consilience has not occurred, those problems have surely not been altered. All of this seems to me to lead to the conclusion that consilience of inductions plays no useful role in confirmation theory at all. To capture the significance of consilience of inductions, we need a new theory of theories, and a more precise conception of the role that consiliences play in the history of science.

III. It is well known that for Whewell each induction is theoretical. Even single inductions involve the imposition of a new idea on the data.

As I argued in my earlier paper, this view leaves Whewell with a permanent and deliberate blurring of the distinction between observation languages and theories. On the question of the role that consilience plays in acceptance of theories, this important point about Whewellian induction has not been sufficiently noted, except by Mary Hesse, whose explication of consilience depends heavily upon her recognition of the crucial fact that the theories with respect to which laws become consilient are themselves imposing new ideas on the laws, just as in any other case of Whewellian induction. Her way of putting it is to say that theories pick out relevant analogies between the data subsumed by different laws; I will now try to show that something like her point is indeed the crucial one in trying to understand consilience.

Recall that consilience takes place when two laws thought to be logically independent turn out to cover data classes that at least partially overlap. The question is, How can the transition, from construing two laws as independent to realizing that they are not, take place? Surely if at some given time L_1 and L_2 are thought to be independent, then there *must be good reasons* for supposing that their separate evidence classes are disjoint; there must be good reasons for supposing that two separate theories are required in order to explain the two laws. It is easy to see that two such thought-to-be independent laws can only come to be regarded as nonindependent if some kind of tampering takes place. I suggest that in cases of consilience the tampering is *semantic,* that is, some change in the meanings of key terms in at least one of the laws must take place. A new theory T involves an introduced interpretive idea; thus, two laws L_1 and L_2 can only become consilient with respect to T because T *alters the meaning* of some of the key terms in one or both of the laws, where such alteration renders the laws nonindependent because they now share some reinterpreted terms in common relative to T. In Whewellian terms, at the historically later time t_2, T can deductively entail L_1 and L_2 only because basic terms of L_1 and L_2, and perhaps also some of their background conditions, have been altered in the light of T. Some might even want to say that the *observational bases* of L_1 and L_2 have been altered by the introduction of T, thus also altering key terms used in describing the events described by E_1 and E_2.

In the analysis thus far given, I have been using the term "independence" in a rather special way. Certainly there is no requirement that in a case of consilience the two laws involved be *logically* dependent. Strictly speaking, the two laws (before consilience) are independent only because they are thought to explain disjoint evidence classes. After consilience the laws are not logically dependent; they simply explain some part of one another's evidence. In this sense they are now seen to be *non*independent, as argued above. It might be said that following consilience the two laws are "semantically nonindependent," capturing the idea that if they are to entail the same evidence statements, they must share some semantic content.

Whewell confuses the real point about consilience with his apparent insistence that scientific theories always be spelled out in deductive form. However, by means of the confusions of his own methodology, he himself realized that mere deductive form will not do. Thus he writes:

> But when we say that the more general proposition *includes* the several more particular ones, we must recollect what has before been said, that these particulars form the general truth, not by being merely enumerated and added together, but by being seen *in a new light.*[13]

An even more instructive quote is the following:

> In Induction . . . , besides mere collection of particulars, there is always a *new conception,* a principle of connexion and unity, supplied by the mind, and superinduced upon the particulars. There is not merely a juxta-position of materials, *by which the new proposition contains all that its component parts contained; but also a formative act exerted by the understanding, so that these materials are contained in a new shape.* . . . Our Inductive Tables, although they represent the elements and the order of these Inductive steps, do not fully represent the whole signification of the process in each case.[14]

As Laudan correctly points out, it is during the kind of historical period that Whewell calls an "inductive epoch" that the most prominent cases of consilience come forward.[15] An "inductive epoch" is a period in the history of science when something like what we would now call a scientific revolution has taken place. Maximum theoretical unification has occurred; domains of data thought to be recalcitrant compared with earlier explanations are finally reduced and caught in explanatory nets; the whole character of a given science is so completely and persuasively transformed that scientists working during that period take the problems of science to involve merely "mopping-up" exercises, exercises that take the predictions of already established laws to greater numbers of decimal places. The laws may need refinement; the question of their truth is foregone.[16] But notice that it is precisely during such a period that the most severe alterations in meanings of theoretical terms occur. Whewell's own most frequently cited examples of major conceptual changes marked by consilience are the Newtonian synthesis and what Whewell took to be the victory for the undulatory theory of light. I will stay with his examples in order to show how the phenomenon of consilience is marked in each case by indispensable semantic or conceptual changes.

Whewell, along with countless others, took Newton's inverse-square law as explaining (entailing) at least the following other laws: Kepler's third law, of the proportionality of the cubes of the distance to the squares of the periodic times of the planets; and Kepler's first and second laws, of the elliptical motion of each planet. In addition, Newton's law of the force of universal gravitation explained other apparently dissimilar phenomena, for example, the precession of the equinoxes and the motions of the tides. Now all of the phenomena thus swept under the covering blanket of the inverse-square law were originally expressed in domains of evidence thought to be disjoint. It appears from some of Whewell's own manners of expression, and from many other historical writers as well, that Newton's great achievement was the *deduction* of the laws describing these other phenomena from the inverse-square law. But clearly this deduction cannot have come off in any straightforward way at all, given that at earlier periods the laws involved were thought to be independent.[17] What was required in order for the deductions to work (if indeed, it is finally worthwhile to try to have them work) were severe changes in the terminology of the laws, in the forms of mathematics used for expressing the laws, and in general, in ways of *viewing* the phenomena, to use Whewell's favorite expression. Another, now unpopular, way of expressing the change is to assert that the new overriding theory refers the other domains of evidence to the "same cause." I will return to this point in a moment.

Whewell's second example of maximum consilience in the history of science is the undulatory theory of light. It is the concept of polarization of light that finally consiliates the apparently disparate data. The consilience takes place, however, only on the supposition that undulations are transverse.[18] On this supposition, A. J. Fresnel's great mathematical induction linked the phenomenon of polarization of crystals and the facts of double refraction. Indeed, Whewell even went so far as to try to show that the particle theory of light was eliminable on the assumption of the transverse character of the movement of light. He writes:

> The phenomena of polarized light show that the fits or undulations must have a *transverse* character: and there is no reason why emitted rays should not be subject to *fits* of *transverse* modification as well as to any other fits. In short, we may add to the emitted rays of the one theory, all the properties which belong to the undulations of the other, and thus account for all the phenomena on the emission theory; with this limitation only, that the emission will have no share in the explanation, and the undulations will have the whole.[19]

Surely this is a particularly clear case of the kind of semantic change that is required in order to achieve consilience of inductions. That the undulations must be transverse is not part of the data, but part of the meaning of the term "undulation." If, therefore, we alter the meaning of the term "fit of emission," to include "transverse character of fits of emission," we have achieved exactly that kind of reduction that allows one theory to capture the data of another. For given this critical semantic change (put alternatively, having reduced two sets of distinct data to the same cause), we can now see that all the data of the particle theory are explainable by the undulatory theory. The success of the newly generated theory, moreover, is not an entirely empirical matter; for what seems to count most in cases of consilience is the increase in the generality of a given theory, and a simplification of that theory.

As I pointed out earlier, Mary Hesse has recognized this crucial aspect of Whewellian consilience. For in Whewell's philosophy of science, to regard a theory *T* as achieving consilience involves more than deductively deriving other laws from *T;* in addition it involves that *T* must show that the apparently different systems covered by certain laws are "in fact diverse manifestations of the same cause."[20] I have written elsewhere about Whewell's treatment of the relationship between theories and causes; a repetition of some features of that discussion may help at this point.[21]

This essay, like most recent ones on the concept of consilience, has stressed Whewell's notion that in consilience different *kinds* of data come to fall under the same explanatory theory. But Whewell used an alternative vocabulary in discussing consilience, and this alternative use is especially interesting in connection with his discussion of Newton's methodology in *On the Logic of Discovery.*[22] In discussing Newton's first rule of philosophizing—"We are not to admit other causes of natural things than such as both are true, and suffice for explaining their phenomena"—Whewell links the notion of *vera causa* with his own concept of consilience. He writes:

> When the explanation of two kinds of phenomena, distinct, and not apparently connected, leads to the same cause, such a coincidence does give a reality to the cause, which it has not while it merely accounts for those appearances which suggested the supposition. . . .
>
> When such a convergence of two trains of induction points to the same spot, we can no longer suspect that we are wrong. Such an accumulation of proof really persuades us that we have to do with a *vera causa.*[23]

The interesting thing about Whewell's talk about the *vera causa* is not that he thought that we can somehow—quite independently of theory—identify causes. Exactly the contrary is the truth. For Whewell, covering distinct and thought-to-be-different classes of data by the same law *is equivalent to* discovering a true cause.[24] What *is* interesting about Whewell's discussion of Newton's first rule in terms of discovering true causes by means of consilience arguments is that it gives us some more adequate idea of just what kind of measure of the truth (or acceptability) of a theory Whewell thought consilience to be. Indeed, his talk about the *vera causa* also tells us something about the role that he thought *theories* play in induction. The force of universal gravitation is a true cause, transverse undulations are true causes, because both are concepts ingredient in theories that achieve, via these concepts, large reductions in empirical subject matters. They function in theories that unify and simplify large accumulations of apparently distinct data. In important general theories such causal concepts supply a new semantics in terms of which lower-level laws are reinterpreted, and domains of data are seen not to be really disjoint at all. But the usual logical and epistemological point has at just this point been misplaced. Any theory *T* that entails a number of laws $L_1, L_2, L_3 \ldots$ is confirmed just insofar as its entailed laws are confirmed. If the *L*'s involved are all true entailments of *T,* then they already possess enough common semantic content provided by *T,* and consilience as a historical happening may be important as a measure of *T* as compared with *other competing T's;*[25] but consilience does not provide any additional confirmation of *T* that would not have been forthcoming from any of *T*'s entailed *L*'s taken in conjunction.

IV. It should be remembered that by the time we come to deal with theories of sufficient generality to talk about the consiliences they achieve, we are already dealing with theories whose success in prediction and explanation is very marked, and here the history of a given science has much to say to us concerning questions of epistemological acceptability of theories. As theories become more general—and for Whewell the successive generalization achieved by a theory is the key mark of its acceptability[26]—they tend to exhibit two properties, consilience and simplicity. Indeed, Whewell suggests that these two properties are identical, or at least that they are "hardly different; they are exemplified by the same cases."[27] Whewell was perfectly well aware of some of the logical problems of the hypothetico-deductive methodology that he developed. He was aware that acceptance of theories is not just a matter of counting up confirmed instances entailed by general statements; by parity of reasoning, I think that he would have accepted that no amount of Popperian corroboration (success of a general theory in standing up to repeated hard tests) could lead to final acceptability of a theory. He does insist that theories be put to the test (though his concept of what experimentation shows is oddly rationalistic and less concerned with taking empirical matters seriously than is common);[28] but at the same time he often speaks about induction as "beyond the reach of method" and even denies that it is a kind of reasoning at all.

Despite his lack of care in logical matters, and despite the excessive oddness of much of his discussion of induction, I nevertheless think that the heart of his discussion of methodology can be saved and reconstructed partly along lines suggested in the essay by Salmon mentioned earlier. To begin, I think we must try to take seriously two things that Whewell says he was attempting to do, even though the two things seem incompatible to many philosophers of science. Whewell *was* trying to provide a methodology of discovery, and he also *was* trying to show how, even in the absence of a logic of induction with the kind of integrity possessed by deductive logic, general theories get to be justified or become acceptable. The interesting thing to note is that these two attempts come together at a common point, because it turns out that consilience and simplicity of theories, taken as marks of the acceptability of those theories, also play a role in suggesting *the kind of science that should be done.* Whewell admitted that there are no rules of method which, once learned, will automatically generate acceptable inductive generalizations. He refers constantly to the role of invention, genius, and subjective guessing in the formulation of new hypotheses in the history of science. But these admissions do not commit Whewell to the conclusion that successful scientific theories teach us *nothing* about how science ought to be done. It is, of course, possible to manufacture new hypotheses at will—everyone knows that. But on any view of science (except perhaps that of P. K. Feyerabend), the known cases of previously successful scientific hypothesizing ought to count for *something.* At the very least, we ought to seek new hypotheses that are relevantly similar to those which have been successful in the past. Total systems exhibiting large measures of consilience and simplicity are known to have been successful in the past; therefore, seek systems of hypotheses that will exhibit consilience and simplicity. Such a methodological syllogism and its resulting rule may seem not to prohibit anything. But this is not the case. As we have seen, true cases of consilience involve semantic changes of a fundamental kind. The theories that we find acceptable and worthy of imitation cannot be mere conjunctions of hypotheses; the systems involved must have reference to a single cause, an overarching generalization that has reinterpreted all of the lower-level laws.

The view of science at issue here is essentially conservative. I am not sure how it could be otherwise. For those who claim that science should have epistemological priority over other forms of knowledge claims are, in effect, championing a certain view of the presumed *rationality* of science. (Notice, I am here talking about various competitors in the game of knowledge production; I am not suggesting that there are no irrational or nonrational aspects in the scientific enterprise.) And that view of rationality, Whewell was suggesting, comes to this: *Bet on past winners.* Consilience is a clear property of those scientific systems known to be successful: Imitate those systems. That is Whewell's rule of discovery; if you will pardon the phrase, his "logic" of discovery.

However—as is usual in such delicate matters—a distinction must be drawn. Whewell is not saying that one must repeat the same science over and over again merely in different languages; discovery is not translation. He is only suggesting, on my present reading of his view of discovery, that we try out the same *kinds* of hypotheses that have been successful in the past. And this suggestion is clearly consistent with thinking that there are no rules for generating *particular* hypotheses that will pass all of the acceptable tests. In matters of discovery, we seem to have two extreme choices (at this point I leave the middle-level choices for others to ponder). In the selection of new hypotheses to be investigated, we either accept some guidance, or we accept no guidance at all. The choice seems to be between Feyerabend's "Anything goes" and Whewell's "If *anything* goes, nothing goes." But if we are to give an explication of science along the lines proposed by Whewell ("MAN is the Interpreter of Nature, Science the right interpretation"),[29] then we had better put our money on what scientists have taken to be the most successful theories and try to find some element common to all those theories that is worth trying to imitate. Consilience, or simplicity, is just that feature of successful science that Whewell thought to be the key to methodological recommendations having to do with attempted discoveries. And one insightful and perhaps productive way of putting Whewell's point is Salmon's suggestion that prior probabilities of introduced hypotheses are *plausibilities.* Along Feyerabendian lines, one might want to suggest that we introduce the most *implausible* hypotheses possible, and then see what happens. But Feyerabend himself knows that there is more to the story than this. What, after all, is the difference between talking about "natural interpretations" and talking about consiliences brought about through conceptual changes in laws?

For Whewell, when talking about possibilities of discovery, the matter at issue has to do with entertaining hypotheses enough like previously successful ones that there seem to be good enough odds that rationality will be preserved. It is like saying that with respect to scientific systems, it is not so much a matter that the moves be truth-preserving, as that they be rationality-preserving. Whewell's view here is unambiguous. Each science, he thought, took its ontological and epistemological start in what he called a "Fundamental Idea." Such Ideas are the ideas of Space, Time, and Number. Whewellian Ideas function something like the categories of Kant, though Whewell is in no sense as precise as Kant in his discussion of these most basic of all concepts. In any case, to have a genuine logic of discovery, that is, a logic that generates decision procedures analogous to those available in deductive logic, one would at the very least have to provide some algorithm for generating Fundamental Ideas. Whewell simply flatly denies that there is any such procedure. What he says is important for our present purposes:

> Although, as we have said, we can give few precise directions for this cardinal process, the Selection of the Idea, in speculating on phenomena, yet there is one Rule which may have its use: it is this:—*The idea and the facts must be homogeneous:* the elementary Conceptions, into which the facts have been decomposed, must be of the same nature as the Idea by which we attempt to collect them into laws. Thus, if facts have been observed and measured by reference to space, they must be bound together by the idea of space: if we would obtain a knowledge of mechanical forces in the solar system, we must observe mechanical phenomena.

Kepler erred against this rule in his attempts at obtaining physical laws of the system; for the facts which he took were the *velocities,* not *the changes of velocity,* which are really the mechanical facts. Again, there has been a transgression of this Rule committed by all chemical philosophers who have attempted to assign the relative position of the elementary particles of bodies in their component molecules. For their purpose has been to discover the *relations* of the particles *in space;* and yet they have neglected the only facts in the constitution of bodies which have a reference to space—namely, *crystalline form,* and *optical properties.* No progress can be made in the theory of the elementary structure of bodies, without making these classes of facts the main basis of our speculations.[30]

Making allowances for the curious nature of Whewell's second example, the crucial point of his supposed rule is clear enough. He appears to me to be saying that no newly introduced hypothesis can be regarded as rationally entertainable if it involves, for its support, either *ad hoc* hypotheses (this restriction is apparently involved in the very idea of consilience), or propositions referring to *a different logical order of things* than those constituting the denotation of the new hypothesis. On one way of reading his rule it is trivial: If you introduce a proposition about *x*'s, don't suddenly regard it as talking about *y*'s. (Alternatively, if you introduce a hypothesis about *x*'s, make quite sure it is talking about *x*'s, and not about *y*'s—after all, you are the one who is responsible for picking the domain of its application.) But there is a second, nontrivial, reading of the so-called rule. On this second reading the rule enjoins consistency in application, as well as adherence to already successful modes of explanation. Whewell would say to Kepler: "Kepler, you are on the right track, because you are adopting mechanical forms of explanation, and such forms of explanation are already known to be successful. Indeed, you might be on the way to a grand synthesis of mechanical phenomena, if you only will stick to your own vocabulary. If you do not stick to this already well-attested vocabulary, you will reduce the rationality of your efforts by complicating the number of factors that we have to take into account. Others will call your results inconsistent; I will report that you were not doing science, but something else, the credentials of which are not very persuasive." Whewell—and I hope today's philosophers of science work in the same context—was interested in discovering what science was all about, largely because he thought that science was the right, true, or more acceptable form of explanation of how things work. His rule, therefore, enjoins us to build success upon prior success, or at least to expect success in speaking a language that others have successfully spoken.

Earlier in this essay I endeavored to show two things: (1) that consilience, regarded as a *measure* of the confirmation of a theory or law, adds nothing to the ways in which the standard accounts of confirmation permit us to assess confirmation strengths; and (2) that consilience, regarded as a strong feature of scientific systems that are deductive in form, must be read as a kind of license to overhaul the semantic interpretations of whole systems of laws. Both points depend upon the model of consilience that I sketched in section II. We are now in a position to see that that model generates extreme difficulties for explicating the concept of consilience in such a way that this concept can be salvaged. If point (1) is right, then consilience cannot operate as a measure of confirmation strength of a system; point (2) seems simply not to fit cases of actual science. For surely many examples can be produced of laws ingredient in highly consilient systems (for example, the Newtonian synthesis) which do not pick out *the same* evidence classes, although there remains a sense in which we can say that the laws are "about the same things."[31] No amount of semantic tampering would seem to allow us to get Hooke's law and Boyle's law to range over the same evidence class, but the two laws are about the same kind of thing, namely properties of elastic bodies. My model thus seems too narrowly conceived to do the required job. The reasons for the major stresses on this model have to do, I think, with the fact that one must initially attempt to explicate consilience in the context of a hypothetico-deductive account of the structure of science.

What I am now suggesting is that we make a major shift in our thinking about consilience consistent with what Whewell wanted in the way of a methodology of science. It is empirically possible to determine the frequency of success of certain types of explanation (Whewell would probably accept this as one of the consequences of his view that the *philosophy* of science is a philosophy of a certain *history* of science), and it turns out that successful types of hypothesizing occur in systems having the property of consilience. When Whewell says that consilience is a test of truth, I think that we should read him as saying: "Look at the history of science; the successful ways of proceeding are those that generate systems of maximal content and which are simple; I call such systems 'consilient' and recommend that science achieve rationality preservation by seeking more such systems." De Morgan was right, Whewell had no inductive *logic;* the *methodology* developed in Book III of *Novum Organon Renovatum* involves an empirical collection of rule-governed ways of doing what Whewell thought of as "standard," that is, successful, science.

In thus suggesting that consilience be viewed as something like one of Salmon's plausibilities, I am not suggesting that the Bayesian reconstruction of scientific inference is necessarily the right one. What I am suggesting is that Whewell and Salmon both realized that the historical success of scientific theories is relevant to the kinds of science that we ought to try to do. Working in this way, we can at least know *what it is that we would have to abandon* in the face of a challenge from a newly recommended methodology. And if accepting the new ways of procedure forced too many readjustments in our methodology and its attendant ontology, we would have ways of arguing that the challenger was less rational, less the "right interpretation of nature."

There is nothing in all of this that prohibits the scientist from following Darwin's recommendation that every scientist be allowed one "damned-fool experiment" (he blew a trumpet at a bed of tulips) on the off chance that something interesting might turn up. What is at issue is finding some guidance in the matter of choosing between competing kinds of hypothesizing. After all, it is explanation, not the discovery of surprises, that science ultimately seeks.

NOTES

1. See the following papers in I. Lakatos, ed., *The Problem of Inductive Logic* (Amsterdam: North Holland, 1968): M. B. Hesse, "Consilience of Inductions," pp. 232–46; L. J. Cohen, "An Argument that Confirmation Functors for Consilience Are Empirical Hypotheses," pp. 247–50; J. L. Mackie, "A Simple Model of Consilience," pp. 250–53; W. C. Kneale, "Requirements for Consilience," pp. 253–54; and M. B. Hesse, "Reply," pp. 254–57. See also Larry Laudan, "William Whewell on the Consilience of Inductions," *The Monist,* 55, no. 3 (1971):368–91; Mary Hesse, "Whewell's Consilience of Inductions and Predictions," ibid., pp. 520–24; and Larry Laudan, "Reply to Mary Hesse," ibid., p. 525. My paper, "Whewell's Logic of Induction," is in *Foundations of Scientific Method: The Nineteenth Century,* ed. R. Giere and R. Westfall (Bloomington: Indiana University Press, 1973), pp. 53–85. I have rewritten portions of the present essay, largely on the basis of comments made by Adolf Grünbaum and Larry Laudan. I also learned a lot about Whewell's methodology from students in my spring 1971 seminar on nineteenth-century British methodology, especially Brian Cupples, Parker English, and Danny Steinberg.
2. Both Augustus DeMorgan and John Stuart Mill were particularly apt at locating Whewell's logical lapses. See, for example, the brief discussion of DeMorgan's criticisms in Robert E. Butts, ed., *William Whewell's Theory of Scientific Method* (Pittsburgh: University of Pittsburgh Press, 1968), pp. 24–26.
3. Wesley C. Salmon, "Bayes's Theorem and the History of Science," in *Historical and Philosophical Perspectives of Science,* ed. Roger Stuewer (Minneapolis: University of Minnesota Press, 1970), p. 77, contains a good brief survey of difficulties with the hypothetico-deductive account. The hypothetico-deductive account presupposes a firm distinction between theoretical and observational sentences. This distinction has been called into question by a number of writers for a number of reasons, and if it is true that all observation is theory-laden, the hypothetico-deductive method is in trouble on that score as well.
4. Salmon, "Bayes's Theorem," p. 77.
5. Ibid., pp. 85–86.
6. Butts, *William Whewell's Theory,* pp. 138–39.
7. Butts, "Whewell's Logic of Induction," pp. 73–76; and idem, "Whewell on Newton's Rules of Philosophizing," in *The Methodological Heritage of Newton,* ed. R. E. Butts and J. W. Davis (Toronto: University of Toronto Press, and Oxford: Basil Blackwell, 1970), pp. 145–47.
8. "Whewell's Logic of Induction," pp. 57–61.
9. Hesse, "Consilience of Inductions," pp. 239–46.
10. Laudan, "William Whewell on Consilience of Inductions," p. 374.

11. Hesse, "Whewell's Consilience of Inductions and Predictions," p. 520.
12. Ibid., p. 521.
13. Butts, *William Whewell's Theory*, pp. 169–70.
14. Ibid., p. 163. Emphasis added.
15. Laudan, "William Whewell on Consilience of Inductions," pp. 384–87.
16. See my "Whewell's Logic of Induction."
17. Recall that Pierre Duhem had argued that some of the laws involved in the Newtonian synthesis not only did not relate to one another deductively, but were in fact *logically* incompatible (Pierre Duhem, "Physical Theory and Experiment," in *Readings in the Philosophy of Science*, ed. H. Feigl and M. Brodbeck [New York: Appleton-Century-Crofts, 1953], p. 245). An irony in Whewell's own work is the fact that he knew that in the case of the tides, there were not even adequate data for the inverse-square law to describe, until Whewell himself had collected such data in the mid-nineteenth century! Unfortunately, a detailed study of Whewell's work on the tides has not been made. I make brief reference to his work in the entry "William Whewell," *Dictionary of Scientific Biography*, ed. C. C. Gillispie (New York: Scribner's, 1972–1975).
18. Butts, *William Whewell's Theory*, p. 157.
19. Ibid., pp. 260–61.
20. Hesse, "Whewell's Consilience of Inductions and Predictions," p. 523.
21. "Whewell on Newton's Rules of Philosophizing," pp. 139–42.
22. Butts, *William Whewell's Theory*, pp. 332–37.
23. Ibid., pp. 330, 331.
24. Butts, "Whewell on Newton's Rules of Philosophizing," pp. 139–40.
25. I will be suggesting shortly that the importance of consilience as a measure for comparing competitive theories (including methodologies) has been largely unnoticed, and that without such comparative theoretical situations, consilience cannot be taken as a measure of the acceptability of theories at all. Like simplicity as usually viewed, consilience only takes on importance as a mark by means of which, when comparing theories, we take one to be preferable to some other.
26. Butts, *William Whewell's Theory*, pp. 160 ff.
27. Ibid., p. 159.
28. See my "Whewell's Logic of Induction."
29. Whewell, *Novum Organon Renovatum* (London, 1858), p. 5.
30. Butts, *William Whewell's Theory*, pp. 216–17.
31. I am indebted to Larry Laudan for providing criticisms that convinced me that my model for consilience was too narrow. I believe that there is still much to be done on the question of when two hypotheses are about things "of the same kind," but I now see that sameness does not necessarily mean membership in the same evidence class.

MICHAEL RUSE
University of Guelph

Is Biology Different from Physics?

> These studies, which now are far advanced, have brought to light, in the objective domain, so many data of both human and animal economy that we may safely say that consciousness emerges within behavior as a new pattern of performance in much the same way that life emerges from the non-living and that every other novel pattern arises by directive action intrinsic to the appropriately organized structure. The distinctive properties of our so-called spiritual life arise within us as naturally as do our breathing and our muscular contractions. It is as natural for the brain to think as it is for the heart to beat.
>
> —C. Judson Herrick
> *"Integrative Levels in Biology," 1949*

Let me begin this essay by admitting to a prejudice—or, at least, to a bias. The biologists I know, researchers, teachers, and students, seem all to be fairly normal kinds of people—as normal, that is, as the members of any other group of people. When off duty, they talk in understandable, unexceptional kinds of ways, they do understandable, unexceptional kinds of things, and they seem to have the same kinds of desires and emotions as the rest of us. This being so, I start off with an *a priori* prejudice or bias toward the question posed in the title of this chapter: All other things being equal, I fully expect biology *not* to be so very different from physics. It seems to me that the presumption is that the biologist, a man with essentially the same kind of thought, language, and criteria as the rest of us, will produce essentially the same kind of scientific work as anyone else, specifically the physicist.

Note, however, that although I confess to this prejudice, I do allow a *ceteris paribus* clause. I do not want to bring the discussion to an end before it really begins, by concluding at once that because biologists are like physicists, biology is and must be the same as physics. I recognize that my prejudice must be tempered by the possibility that all things are not equal for the physicist and biologist—that indeed there are considerations which lead to crucial divergences between physics and biology. But it does seem to me that the initial onus is upon him (or her) who would separate physics and biology to make his (or her) case and to give good reasons why things are not equal for the physicist and biologist, rather than upon me to show why physics and biology are essentially similar.

As anyone with but the slightest acquaintance with the literature knows well, there is no shortage of putative good reasons, all claiming to erect impassable barriers between the physical sciences and the biological sciences. In this essay I shall consider two such reasons, and I shall suggest that they both have some force, but that neither proves all that their proponents seem to think (or have thought).[1] The first reason revolves around the question of *complexity;* the second revolves around the question of *teleology*. I take them in turn.

1. BIOLOGICAL COMPLEXITY

The first argument that biology is in some crucial way different from physics is fairly easily understood.[2] Although there is considerable controversy about the precise nature of physics, more specifically a physical theory, it is generally agreed that such a theory is essentially a system of related laws.[3] To be more exact, such a theory is a "hypothetico-deductive" system, meaning that a number of laws are taken as axiomatic—unproven within the system—and from them all other laws are inferred deductively. By "law" in this context is meant a universal, true statement about matters of fact, thought in some way to be necessary (such physical necessity is usually distinguished from logical necessity by the name "nomic" necessity). The paradigm for this view of scientific theories is, of course, Newtonian dynamics with Kepler's laws being deductively inferable from Newton's laws of motion and gravitation.[4]

Now, so the argument continues, although this kind of theory functions very successfully in the physical world, it flounders when we turn to the biological world. At all levels of biological phenomena we encounter incredible complexity, from the level of the cell right up to the species and beyond to the organic world taken as a whole. Thus, for example, J. D. Watson—certainly no booster of the autonomy of biology—is forced to admit when writing of *E. coli,* one of the simplest of all organisms, that

> [some] cellular molecules, in particular the proteins and nucleic acids are very large, and even today their chemical structures are immensely difficult to unravel. Most of these macromolecules are not being actively studied, since their overwhelming complexity has forced chemists to concentrate on relatively few of

them. Thus we must immediately admit that the structure of a cell will never be understood in the same way as that of water or glucose molecules. Not only will the exact structure of most macromolecules remain unsolved, but their relative locations within cells can be only vaguely known.[5]

And this is but a start. Complexity multiplies as we move up the scale. Just think, for instance, of the incredible number of relations among the members of a group of organisms—to say nothing of the effects of the external environment.

But what is the consequence of all of this complexity? It is this, that it is difficult if not impossible to draw any significant general statements about what occurs in the biological world. As soon as one puts forward any general statement, one runs immediately into one counterexample or exception after another. Hence, one has to weaken one's statement until it becomes practically a truism—"All organisms that are born die," "All organisms show some adaptations for survival and reproduction, although perhaps not for their own survival and reproduction"—or one has to restrict the scope of the statement until it can hardly be called "general" at all. Consequently, one fails entirely to get genuine biological laws—laws that measure up to the criteria set for laws of physics. Moreover, even with the limited generalizations one can get, the incredible complexity of biological phenomena rules out any very successful attempt to connect them systematically; certainly, hopes of a full-blown hypothetico-deductive system seem doomed to disappointment.

Hence, in biology, so the argument goes, we have to settle for rather different kinds of understanding and explanation—more descriptive, more tuned to the particular, more ready to accommodate to the complex. And, as is well-known, there has been a proliferation of suggested models of biological understanding, models which do not demand full-blooded universal laws or tight logical connections between premises and conclusions. Among others we can mention "narrative explanation," "integrating explanation," and "characteristically historical explanation."[6] Depending on the particular ax one is grinding, either biology's reliance on these models is taken to be a mark of its distinctive autonomy (biology is different from physics but an equal partner in the scientific endeavor), or it is taken to be a mark of biology's inferiority (biology fails when judged by the criteria of the best kind of science, that is, physics).

Although there does not seem to be any absolutely logically necessary connection, coupled with the argument about complexity is usually one about uniqueness.[7] It is felt that all of the intricacy of the biological world leads to just about everything being different from everything else. But laws demand repeatability; they cannot be applied to unique phenomena. Therefore, in principle, as well as in practice, there can be no laws (and hence no hypothetico-deductive systems) of biology. Thus, for instance, Ernst Mayr writes: "In the uniqueness of biological entities and phenomena lies one of the major differences between biology and the physical sciences. . . . [In] the organic world . . . all individuals are unique, all stages in the life cycle are unique, all populations are unique, all species and higher categories are unique, all inter-individual contacts are unique, all natural associations of species are unique, and all evolutionary events are unique." And he adds: "It is quite impossible to have, for unique phenomena, general laws like those existing in classical mechanics."[8] In a similar vein, rolling together the arguments about complexity and uniqueness, G. G. Simpson writes that "[in] complex events, it is evident that the extremely intricate configurations involved in and necessary, for example, as antecedents for . . . the origin of *Homo sapiens* simply cannot recur and that there can be no laws of such one-of-a-kind events."[9] And, also putting together complexity and uniqueness, W. M. Elsasser writes: "Organisms are structurally and dynamically so complex that one can always find individual differences. . . between any two organisms of the same class no matter how one defines the class."[10]

Now, as many have pointed out, at one level one can tear these arguments to shreds.[11] Complexity per se is no absolute bar to the discernment of laws, or to the incorporation of these laws into theories. And the whole argument about uniqueness rests on an equivocation: In one sense everything is unique, in another sense nothing is. Every swinging pendulum has its own unique spatiotemporal coordinates, and even Simpson would admit that there were some similarities between the origin of man and the origins of, say, the great apes—for example, they both involved genetic change. But this is not, I think, to deny considerable truth in the arguments—at least the argument about complexity. Biologists do encounter fantastically complex phenomena, and as soon as they seem to have grasped some kind of regularity, a counterexample springs up. Thus, for example, Simpson holds up "Allen's rule," namely that when mammals adapt to colder climates their feet become shorter.[12] This is estimated to have an exception rate of 36 percent! And, it could well be argued, some of the more reliable laws of biology are reliable only because they hover on the edge of analyticity. Take, for instance, Dollo's law, that evolution cannot reverse itself. When one points out that the ancestors of whales were aquatic, then became land animals, and are now back in the sea, one is told that the law is not intended to preclude this. When one points out that some groups nearly achieve reproductive isolation but then hybridize, one is told that the law is not intended to preclude this. When one points out that one gets not only mutation from allele A_1 to allele A_2, but also back mutation from allele A_2 to A_1, one is told that the law is not intended to preclude this—and so on and so on.[13]

But granting all of this, is this still not to take altogether too gloomy a view of the picture? Even if it does not hold true generally, surely there are some parts of biology where one can find laws, laws that are solid enough to bind into theories? Since one of the most distinctive marks of physical theories is the extent to which they use the ideas and techniques of mathematics, perhaps we might most quickly answer our question about biology if we turn our attention to those areas of biology which in a like manner make great use of mathematics. And from these areas, one particular subject stands out, namely *population genetics*. It seems to me that if this discipline fails to manifest characteristics of physical science, then we must truly conclude that biology is not very much like physics.

2. POPULATION GENETICS

Population genetics is concerned with the distributions of the causes of organic characteristics and variations in populations, and the ways and reasons why these distributions vary from generation to generation. The basic cause of organic characteristics and variation, the gene, is thought to be governed by a number of relatively simple laws, chief among which are Mendel's two laws—the law of segregation and the law of independent assortment. Although neither of these laws applies directly to group situations, they can easily be generalized, and thus one obtains (among others) what is probably the most fundamental law in population genetics, the Hardy-Weinberg law. This, in its simplest form, states that in a random-mating, effectively infinite population of sexual organisms, given an alternative pair of alleles, A_1 and A_2, at some particular gene locus, in the absence of external factors like selection and mutation, if the initial ratio of A_1 to A_2 is p to q, then for every succeeding generation the ratio will be p to q, and, no matter what the distribution of genotypes in the initial generation, for all succeeding generations the distribution will be

$$p^2 A_1 A_1 + 2 p q A_1 A_2 + q^2 A_2 A_2$$

Given this law, the population geneticist can introduce a number of complicating factors, such as mutation, selection, and inbreeding, and he can devise sophisticated models explaining and predicting the distribution and spread of genes in infinite and finite populations under many different situations. And, moreover, it is indisputable that the consequences derived in population genetical models follow *deductively* from the premises.[14]

From a formal viewpoint, therefore, population genetics seems very much like a theory of the physical sciences—a statistical theory of the physical sciences, for, from the Hardy-Weinberg law on, everything in population genetics involves ratios and distributions over groups. An immediate analogy that springs to mind is the kinetic theory of gases. Thus, for example, one has in that theory the so-called Maxwell-

Boltzmann distribution. This enables one to calculate the probable speed of molecules in a gas and to show the distribution of the different speeds of molecules in the gas. More specifically, if the function $f(v)$ is understood as the fraction of molecules per unit speed range (the "speed-distribution function"), and v_o is the most probable molecular speed, then the Maxwell-Boltzmann equation states that

$$f(v) = Bv^2 e^{-v^2/v_0^2},$$

where

$$v_0^2 = 2kT/m \qquad \text{and} \qquad B = (4/\sqrt{\pi})\,(1/v_0^3).$$

Then, with the aid of this equation, one can turn to actual cases. Thus, if one has nitrogen at 20°C, the most probable speed (v_o) is calculated as 4.2 $\times$ 10^4 cm/sec, and, for instance, the fraction of molecules in the speed range between 0.9 v_o and 1.1 v_o is about 17 percent.[15]

In a very similar manner, one can develop and apply population genetics. Thus, suppose one has a case of superior heterozygote fitness ("heterosis"). This implies that the heterozygote genotype $A_1 A_2$ is fitter than either homozygote $A_1 A_1$ or $A_2 A_2$. Suppose in particular that the coefficients of selection against $A_1 A_1$ and $A_2 A_2$ are S_1 and S_2 respectively (that is, the chance of some particular $A_1 A_1$ or $A_2 A_2$ surviving and reproducing is only 1-S_1 or 1-S_2 as opposed to the chance of an A_1A_2). It then follows that we shall get a balance between the three genotypes in the population, a balance which is a function of the different selective intensities. In particular, if the proportion of A_1 to A_2 alleles is q to 1-q, then

$$q = S_2/(S_1 + S_2).$$

Applying this to a particular situation, if we suppose that $S_1 = 1/4$ and $S_2 = 1$ (that is, no S_2's at all survive and reproduce), then it follows that $q = 4/5$, and that for every generation, of a typical 100 new members of the population, 64 will be $A_1 A_1$, 32 will be $A_1 A_2$, and 4 will be $A_2 A_2$. In other words $A_2 A_2$'s keep reappearing although they never reproduce.[16]

In neither the gas nor the genetic case can one predict or explain the fate of some individual. What one can predict and explain are the fates of groups of individuals—molecules in the one case and genes in the other—and the examples given above show that the formal reasonings in both the physical and biological cases are very similar. With the aid of the Maxwell-Boltzmann equation one can calculate the percentages of molecules of a certain kind (molecules within a certain speed range); with the aid of the Hardy-Weinberg law one can calculate the percentages of genes and genotypes of a certain kind.

In passing, two points are worth noting. First, because the population geneticist does not deal with genes on an individual basis, this does not imply that he believes their performance to be without reasons or a function of "chance." His refusal or inability to say why a particular allele A_1 (rather than A_2) is passed on from parent to offspring does not imply that he believes there are no reasons, any more than the gas theorist's refusal or inability to predict the speed of a particular molecule implies that he believes the molecule is not subject to the laws of motion. To the contrary, in both cases they believe that there are laws governing the behavior of individuals. It is just that the statistical theories operate at a higher level of generalization, assuming that differences between individuals can be averaged out.

Second, following on the first point, it is a mistake to assume that because population genetics is a statistical theory it is therefore a nondeterministic theory. Let us adopt Nagel's definition of a deterministic system: "The set of laws L constitute a deterministic set of laws for [a system] S relative to [a set of properties] K if, given the state of S at any initial time, the laws L logically determine a unique state of S for any other time."[17] (I assume Nagel here means any *later* time.) Population genetics is certainly not deterministic at the individual level—given parents $A_1 A_2$ the offspring are not uniquely determined, for they could be $A_1 A_1, A_1 A_2, A_2 A_2$—but at the infinite population level it is deterministic, just as is the kinetic theory of gases. Thus, given the state of an infinite population, we can calculate the unique state at another time—for instance, if the ratio of genes is p to q, then the Hardy-Weinberg law shows that future ratios *must* be p to q (in the absence of mutation, selection, and so forth).

However, what must be mentioned is that population geneticists have extended their theory to consider *finite* populations, as well as infinite ones. This, it seems to me, does take them to nondeterministic considerations. Thus, for example, one important result is that if one begins with a finite population with a gene ratio evenly divided between neutrally adaptive alleles A_1 and A_2, the ratio of A_1 to A_2 can be expected to "drift" at random, until eventually one or other of the alleles is eliminated from the population.[18] On the above understanding of "deterministic," it certainly does not seem that the success (or failure) of A_1 as opposed to A_2 is determined by the theory of drift (this is not, however, to say that at this point population genetical theory ceases to be deductive; one can show rigorously the rates at which drift might be expected to act). It should be emphasized that at this point I am writing only of the formal aspects of population genetics. As we shall see shortly, the actual occurrence of drift in any significant way is hotly debated.

Returning now to the direct consideration of the formal structure of population genetics, it has to be admitted that nothing is very rigorous, at least, not by the standards of formal logicians. One does not find a few axioms carefully specified, legitimate rules of inference noted and listed, and all theorems meticulously deduced from preceding parts of the theory. However, as everyone who knows what real science looks like is aware, contrary to the impression given by many elementary books on the philosophy of science, one rarely does find such rigor, even in physics. The population geneticist, just like the physicist, uses unmentioned or semiexplicit mathematical principles, he leaves out "obvious" steps, he reverts to language to illustrate or to underline complex or surprising points, and he often consigns his really difficult deductions to appendices, merely stating essential conclusions in the body of his text.

But despite the looseness of his methodology, it does seem that the axiomatic ideal lies behind the population geneticist's theoretical development of his subject matter. I suggest, therefore, that considered from a formal viewpoint—that is, considered as a more-or-less uninterpreted calculus—population genetics does seem very much like a branch of physics. This now brings me to the question of how it measures up when compared against the real world. Are Mendel's laws really "laws" (understood as judged by the criteria for lawfulness in physics), or are they little more than complex marks on paper, and are all the intricate inferences merely exercises in pure-mathematical ingenuity? As is well known, it cannot be denied that some have tried to dismiss the whole population-genetical corpus at one stroke, arguing that Mendel's laws fall pitifully short of full law status as measured by the canons of physical sciences. Thus, for example, J. J. C. Smart argues that the laws are so riddled with counterexamples that they can be considered as little more than rough rules of thumb.[19] Obviously, if Smart's claim is true, it strikes right through population genetics.

I shall not take time here to counter Smart's argument in detail; this I have done elsewhere.[20] Such plausibility as it has seems to me to stem from an ignorance of biological theory coupled with a preparedness to apply far harsher standards to biology than to physics. By way of indication of the manner in which I think Smart can and should be countered, let me consider briefly Mendel's first law, his law of segregation. This is the law stating that in sexual organisms, every organism gets one of the pair of genes at each locus from one parent and the other of the pair from the other parent, the genes coming from corresponding loci in the parents and it being a matter of chance which of a parental pair gets transmitted. (As emphasized above, the choice of gene is considered "chance" at the level of Mendelian laws; no one thinks there are no reasons or causes at all.) One of the major reasons we elevate an empirical universal statement to law status is, assuming that it does not conflict with other claims to which we afford law status, that it has been found to hold in a wide number and variety of situations, and particularly valued are newfound confirmations which were not, as it were, "built into" the original evidence for or expression of the statement. These latter instances go a great way toward strengthening our conviction that a statement is a law—that is, that it *must* hold. In the light of criteria like this I would suggest that Mendel's first law has as much right to full law status as any of the famous laws of physics. It holds generally in the animal and vegetable worlds, and nearly all its confirmations came after Mendel's first experiments on pea plants. Indeed, the incredible excellence of Mendel's experimental results strongly suggests that Mendel had his law before there were any confirmations.[21]

Conversely, to those who would point to the exceptions to Mendel's first law, the transmissions of extrachromosomal genes, for instance, I would point to the exceptions to physical laws. Snell's law breaks down when confronted with Iceland spar, and Boyle's law is notorious outside a limited range of temperature and pressure. In short, I would suggest that Mendel's first law has sufficient positive marks to count it a genuine law and that its negative attributes are no greater than those of many laws of physics. More generally, I would claim the same of Mendel's second law and of immediately consequent laws like the Hardy-Weinberg law. They have as much right to be called "laws" as famous laws of physics like Boyle's law, Snell's law, and Hooke's law. Hence, it seems to me that, as compared to physics, population genetics does at least start off on the right foot—both in theory and in practice.

But, I think, we have a far greater obstacle to clear than that thrown up by Smart before we can conclude that population genetics *taken as a whole* has all the marks of a mature physical science. It cannot be denied that there is considerable controversy among population geneticists themselves as to the application and significance of major parts of their theory to the real world. Let me explain in a little detail the nature of this controversy and then turn to its philosophical implications.[22]

3. TWO RIVAL HYPOTHESES

There seem to be two major camps into which population geneticists fall. On the one hand there are supporters of what is called the "classical" hypothesis.[23] Essentially they see an interbreeding group of organisms as a collection of similar or near-similar entities. They believe the genotype of each organism to be practically the same as the genotype of every other organism, with nearly every organism being homozygous at any locus with respect to the same allele—the "wild-type." This uniformity is marred only by the occasional mutant gene, usually recessive, nearly always deleterious, and hence kept under close check by selection. Occasionally a mutant proves advantageous, spreads rapidly through the population, and then becomes the wild-type. The opponents of this hypothesis, supporters of the "balance" hypothesis, see genetic diversity everywhere.[24] For them, selection is a mechanism for maintaining difference, not eliminating it. They believe that in an interbreeding group one tends to have several alleles in significant proportions at any locus ("polymorphism") and that organisms consequently are heterozygous at many loci. Diagrammatically, the difference between the two hypotheses can be shown as in figure 1.

Two Different Members of the Same Group

Classical hypothesis	$\dfrac{++++\ldots+m+}{++++\ldots+++}$	$\dfrac{++++\ldots+++}{+m++\ldots+++}$
Balance hypothesis	$\dfrac{A_3B_2C_2DE_5\ldots Z_2}{A_1B_7C_2DE_2\ldots Z_3}$	$\dfrac{A_2B_4C_1DE_2\ldots Z_1}{A_3B_5C_2DE_3\ldots Z_1}$

Note: "+" is a wild type gene; "*m*" is a mutant; "A_1"; etc., are different alleles.

Figure 1

Supporters of the balance hypothesis give several reasons why selection might maintain allelic polymorphism (with the consequent heterozygosis). If selection favors the rarer allele, this will do the trick. As selection gets to work, the rarer becomes less rare until a stable balance is achieved. Such a phenomenon is believed to occur in many predator-prey situations. The predator learns to go for the more common phenotype, and thus the less common gene which causes the less common phenotype has the selective advantage. However, probably the most popular hypothesis for balanced allelic polymorphism is superior heterozygote fitness. As we have seen, if the heterozygote for two alleles is fitter than either homozygote, then this will keep both alleles stably in the population. The classic (critics would say, almost only) example of this centers on sickle-cell anemia, which is maintained in many Negro populations because the heterozygote for the sickle-cell gene has a selective advantage over the non-sickle-cell homozygote owing to improved defense against malaria, even though the sickle-cell homozygote has little or no chance of life.[25]

One might think it would be a comparatively easy matter to decide between these rival hypotheses. Count up the mutant genes and/or heterozygotes in populations and then lay one hypothesis quietly to rest. Classical hypothesis supporters have tried to do just about that. They point out that the number of alleles that are visible mutants (mutants causing unambiguous phenotypic changes) in populations tend to be very small, as their position predicts. They point out also that the number of recessive lethals (alleles which, were they matched homozygously in a carrier, would cause death) in populations also tend to be very small—as their position predicts. Also they ask why, if heterosis is so common, the sickle-cell example keeps behaving like Mrs. Gamp's cucumber—repeating over and over again.

The balance supporters fight back with vigor. They argue that the real stuff of genetic change involves mutations to alleles which have minute differences on phenotypes, and they back up their position by showing how such minute differences can nevertheless be spread by selection. They argue that visible mutants (and lethals), with the gross changes they cause, are atypical and thus neither prove nor disprove either hypothesis. And when asked to provide examples of "typical" differences that different alleles will have on a phenotype, they point to the virtually impossible task of measuring such differences. Thus, an improved reproductive success of but a few percent might give the bearer of the allele causing the success a terrific selective advantage; but the immense fluctuations in reproductive success caused by uncontrollable and unknown environmental conditions make it impossible definitively to pin down any slight improvements in reproduction due to the change from one allele to another.

While all this may be true, one may feel that in arguing this way, balance supporters are defending their hypothesis rather than supporting it. But they do have evidence, strong evidence, in support of their hypothesis. If, as they claim, there is so much variation in every population, then artificial selection should reveal it. And it does indeed. Writing of the population geneticists' favorite organism, the fruit fly, R. C. Lewontin writes: "It is a commonplace of Drosophila population genetics that 'anything can be selected for' in a non-inbred population." Then, having given an impressive list of what in fact has been selected for, Lewontin concludes:

> There appears to be no character—morphogenetic, behavioral, physiological, or cytological—that cannot be selected in Drosophila. The only known failure is the attempt of Maynard Smith and Sondhi in their pattern experiment to select for left-handed flies. They did succeed, however, in increasing the asymmetry, although it was a fluctuating asymmetry not biased toward the right or left.[26]

One assumes that the Ford administration is pouring funds into attempts to bring Maynard Smith and Sondhi's experiment to a successful conclusion.

Obviously these selection experiments strike a blow at the classical hypothesis. Were organisms genetically uniform, then no amount of artificial selection could achieve the changes that have in fact been effected. But the classical hypothesis—or perhaps now more strictly the "neoclassical" hypothesis—continues to thrive, albeit at a different level. Recently, population geneticists have started to employ molecular techniques, with great success. Roughly speaking, the sequence of nucleotides making up a structural gene transcribe into a sequence of amino acids making up a polypeptide chain, and, ignoring redundant nucleotide substitutions, changes in the nucleotide sequence reflect into changes in the amino acid sequence. Several of the amino acids are electrostatically nonneutral; consequently many changes in polypeptide amino acid sequence are accompanied by changes in overall negative or positive charge. These changes are detectable by a technique known as gel electrophoresis, and thus the population geneticist has a way in which he can detect, unambiguously, changes of the most minute kind between one allele and another. Using the polypeptide chains as phenotype, a whole new way is opened to study the gene.

The results are astounding. If one makes allowance for amino acid substitutions which do not involve charge changes, again referring to Lewontin on the fruit fly, "the average heterozygosity per locus for natural populations of *D. pseudoobscura* is about 35 per cent and essentially every gene is polymorphic."[27] Surely, then, this spells the end of the classical hypothesis.

But this is not so. The classical supporter cannot deny this polymorphism, but he continues to deny the balance supporter's claim that this is held in the population by selection. Rather he replies that all, or nearly all, the variation at the polypeptide phenotypic level is of neutral adaptive value; hence, genetic drift is the chief cause of molecular polymorphism.[28]

This may strike the listener as somewhat ungracious *ad hoc*-ism, but the classical supporter gives good reasons why the balance supporter—who sees in the polypeptide variation a true reflection of all organic variation—still faces grave problems. For instance, it can be shown that if an organism is heterozygous at a large number of loci because of heterozygote superior fitness, then simple intuitive models predict that the organism will be ludicrously greater amounts fitter than the average—amounts so ludicrously great, in fact, that they could never be realized in nature. (Lewontin, for instance, generates a *Drosophila* female who ought, theoretically, to lay 10^{43} eggs.) By adding sophistications to his model, like thresholds beyond which fitness cannot rise, the balance theorist avoids this problem, but all his troubles are still not over. He still faces a problem due to inbreeding depression. With much heterozygosity due to heterosis, inbred organisms (which would be high in homozygosity) ought to be very unfit: far more unfit than they in fact appear to be. The obvious way to avoid this conflict between theory and evidence is to assume that there is a lower threshold at work here. Unfortunately, if anything, the experimental evidence (as opposed to what would be theoretically convenient) points the other way.[29]

Finally, to bring this state of the discussion to an end, it should be pointed out that not everything spreads despondency among supporters of the balance hypothesis. Indeed, the balance theorist can point to much new evidence revealed by the study of polypeptides to support his position. The geographical evidence is very strong. If the classical (or neoclassical) hypothesis is correct, then since polypeptide change is a function of drift, we ought to find no similarities in ratios of alleles at different loci between organisms of separate, reproductively isolated groups. On the other hand, if the balance theory is correct, since these ratios are understood as a function of adaptive value (whatever that function may be), we would expect to find many significant similarities. We would expect also to find a few loci where allelic frequencies were quite different between groups. This would be due to a variety of reasons like different environmental (and hence selective) conditions. The evidence strongly supports the balance hypothesis. For instance, in a very recent study on five *Drosophila* species (the *D. willistoni* group), Ayala and coworkers found that "the same alleles, and in similar frequencies, occur in different populations at most loci in all five species."[30] Moreover, Ayala was able entirely to defuse the counterploy that the supporters of the classical hypothesis might make, namely that this similarity was a function of migration between groups. He pointed out that some intraspecific groups are so isolated geographically that the possibility of migration may be discounted, but nevertheless they share the genetic similarities. In an analogous manner, different species, things between which by definition gene exchange is impossible, share the genetic similarities. And conversely, at some loci genetic frequencies are very different in different populations. Were migration the cause of genetic similarity between populations, such differences ought not to occur. In short, Ayala felt able to conclude as follows:

> The predictions derived from the hypothesis of adaptive neutrality of protein polymorphisms are at complete variance with the empirical observations in the *D. willistoni* group of species. We believe that our results place the neutrality hypothesis in serious jeopardy.[31]

4. IS POPULATION GENETICS REALLY LIKE PHYSICS?

Now, what philosophical juice can we extract from this scientific discussion? It is obvious that if what we seek in population genetics is a fully articulated hypothetico-deductive system, the truth and applicability of which are accepted beyond doubt, then we are doomed to disappointment. There is too much controversy for that—although note incidentally that from a formal viewpoint much of the controversy is probably properly interpreted as being a controversy, not about the truth of population-genetical theory itself, but rather about applicability. Neither supporters of the balance hypothesis nor supporters of the classical hypothesis question the truth, for instance, of Mendel's laws, nor (by and large) do they question each other's models as such. Much of the debate is about existential claims—initial conditions and conclusions. Thus, no one denies that if certain rates of mutation, selection, inbreeding, and so on prevail, then certain consequences obtain. The debate is to a large extent about precisely what rates of mutation, selection, and so on prevail, and precisely what consequences obtain. Does one, for example, have consequences that suggest drift?

One might, of course, argue that the different initial conditions supposed by the rival classical (or neoclassical) and balance supporters are themselves consequences of different lawlike hypotheses (together with certain other initial conditions). Thus, the classical and balance supporters hold the two hypotheses: "Selection in a population works in great part toward the elimination of variation" and "Selection in a population works in great part toward the preservation of variation" respectively. Combined with existential assumptions about the existence of populations and assumptions about the ways in which selection might eliminate or preserve variation, one can then generate the different initial conditions that different population geneticists want to feed into their theory. However, although it is convenient to think of different population geneticists as being supporters of different hypotheses (where these are understood as being lawlike claims), in actual practice population geneticists seem to begin their formal calculations by jumping straight to the different initial conditions that different hypotheses would generate. Hence, in actual practice the debate between population geneticists seems to be about which models actually apply in nature (where "model" in this context is understood as a part of the theory with special assumptions pertaining to special situations). Does one, for example, need to use a model postulating very low selection pressures, or does one need a model postulating relatively high selection pressures?

But, accepting the existence of this controversy, where does this leave us vis-à-vis our major question as to whether biology, as personified by population genetics, is like physics? There seem to be two basic replies open, one's choice being dictated by how one views the scientific situation. On the one hand one might feel that, despite the controversy, the major outlines of the true situation are now becoming apparent. This, I must confess, is my own feeling with respect to the balance hypothesis. It seems to me that the evidence, particularly that of artificial selection experiments and of geographical variation, shows in a very strong manner that the final resolution of the controversy will be either a decisive victory in favor of the balance hypothesis or a synthesis in which major elements of the balance hypothesis will figure prominently. Moreover, while it is indeed true that there is evidence against the balance hypothesis in its present form, such evidence does not yet seem to be of such a devastating kind that the hypothesis will have to be or ought presently to be abandoned entirely. For example, although the known evidence from inbreeding points away from rather than toward the predictions of the balance hypothesis, the total amount of evidence on the subject so far accumulated hardly seems sufficient to make a definitive judgment at present one way or the other.[32]

If what I claim about the balance hypothesis is so, or if conversely one feels that the classical (or neoclassical) hypothesis has won the day, then it seems proper to conclude in a fairly positive manner that biology (that is, population genetics) is indeed like physics. Formally the discipline is like a physical theory, it starts off with statements like the Hardy-Weinberg law, which have as much right to be called "laws" as laws of the physical sciences (and which are called "laws" for the same reasons as those of the physical sciences), and now the theory generally is relating to the evidence in a way typical of the physical sciences.

This is not to deny that in many respects population genetics has more the attributes of a research program than of a fully articulated, mature theory of the physical sciences. Apart from anomalies which need to be resolved (although in this context it is salutary to remember how long some of the great theories of physics, like Newtonian astronomy, lived with anomalies), much, much, work yet remains to be done. For example, even if one agrees that the balance hypothesis is basically true, much more needs to be known about the ways in which selection balances different alleles. For instance, is most of the balance due to heterosis, and if it is indeed so due, why are the heterozygotes so often superior to their homozygotes? Can one give general answers to questions like these, or must each case be taken individually? But while population genetics may still be adolescent, it does seem that on this interpretation of the scientific situation, one can properly answer that biology—some biology—is indeed like physics.

What happens, however, if one feels that, as yet, no decision at all can be made about the relative merits and ultimate hopes of success of the balance and classical hypotheses? It is clear that far less can be claimed by he who would argue that biology is like physics. Some claims can be made. One can point to the elementary parts of population genetics, where one finds laws and theory development just as one finds in the physical sciences. And one can cite as successes such things as the explanation of the balance of sickle-cell anemia. So the absolute logical or even practical impossibility of any part of biology ever being like physics seems ruled out. But when this is said, faith and hope have to take over. At best one can argue that quite probably some day substantial parts of biology will be like mature physical theories.

But, it might be asked, does one even have the right to argue this? I would suggest that one does—or at least I would suggest that it would be most unwise to argue the opposite. For a start, even a little bit of success surely justifies some degree of optimism, and for a second, these are absurdly early days to argue that all is hopeless. It is only about ten years since molecular techniques opened up a whole new range of evidence for population geneticists. It would be ridiculously premature to argue already that there is just no way that population geneticists are going to bring their formal calculi and the empirical evidence into harmony (after the fashion of the physical sciences).

It might be suggested that this argument of mine is a classic case of the fallacy of biased statistics: that although these are early days to argue that biology cannot be like physics if we consider only molecular population genetics, these are hardly early days if we consider all biology and that it is all biology we should consider. The problem is whether optimism is justified in the light of the fact that *all* biology, being still controversial within, gives no definitive answer either way. In reply I suggest that population genetics is a special case and that therefore one is justified in treating it in isolation, as I am doing. It is one area of biology which is formalized—a very necessary even if not sufficient step in the direction of being like physics. It has some success in being completely like physics, and until recently the major barrier to making it more like physics was the methodological impossibility of bringing most of the empirical evidence under measurable control. It was not so much that the evidence was too complex to subsume under law; it was rather that one could not get the evidence at all. Now the evidence is starting to flow in, and not surprisingly the previously mainly *a priori* theory is having some difficulty coping with it. But common sense dictates, even if charity does not, that population geneticists should be given some time to bring their formalisms and the evidence into harmony.

Possibly one might argue that even if it were logically and practically possible that the empirical evidence someday be brought under a theory like one of the physical sciences, given population geneticists' methodology it is highly unlikely in fact that such an event will occur. But this seems a poor argument. If anything, the evidence points the other way. As one studies the controversy in population genetics, one starts to get a strong sense that the methodology is all very familiar. Population geneticists do not turn to fancy modes of explanation, like narrative explanation, designed to capture the peculiar complexity or uniqueness of biological phenomena. Rather, both balance and classical supporters ride roughshod over the supposed uniqueness problem, both blithely assuming that instances of particular forms of genes, alleles, are repeatable time and again—repeats which all assume involve absolutely identical copies of the same thing. And all try to relate their formal theory to the facts in the way a physical scientist would: The problem is not a proliferation of non-physical-type explanations but a proliferation of rival physical-type explanations. This holds, I would add, even for those who would say to both classical and balance hypothesis supporters, "A plague on both your houses." Lewontin, for instance, falls into this camp, but his third alternative is supposedly better because it alone yields (or will yield) true predictions.[33]

In short, I would contend that it is far too early yet to argue that population genetics will never be like a physical science theory, that even the most pessimistic must allow that some success has been achieved and that therefore one has some good grounds for optimism that more success will be achieved, and that if success is not achieved it will certainly not be because population geneticists have deliberately turned their backs on the methodology of the physical sciences. Finally, I would point out that if one is determined to tear apart physics and biology, and if one (unreasonably in my opinion) argues that the present situation in population genetics points to the impossibility of a fully developed biological theory ever being like a physical theory, then one still has to live with and acknowledge the limited success of population genetics. Moreover, one must recognize that while biology may be peculiarly complex, one ought not to rely for one's claims on the supposed uniqueness of biological phenomena. To the contrary, one is arguing that biology is not like physics precisely because biological phenomena are not unique! Part of one's case, for example, rests on the fact that there is a clash between the assumption, on the one hand, of widespread heterosis (thus explaining geographical variation) and, on the other hand, the lack of predicted inbreeding depression. If this were something that happened once, in one group, population geneticists might be prepared to shunt it to one side, as an inexplicable anomaly. The trouble is, of course, that it is believed to happen time and again, and because of this the defender of biological autonomy feels that he can point to the impossibility of a biological theory being like a physical theory. In other words, the appeal to complexity rules out an appeal to uniqueness.

5. BUT IS POPULATION GENETICS REALLY BIOLOGICAL?

There is perhaps one possible counterobjection I should consider before continuing. It might readily be agreed that population genetics is a theory in fact or potentially *like* the physical sciences; but this, so the objection would go, is hardly any surprise, since population genetics is a theory *of* the physical sciences. Now that the population geneticist concerns himself with polypeptide chains and regards changes in these chains as reflections of changes in the DNA sequences of organisms, it is hardly still proper to think of population genetics as a theory of biology. It is now a theory of the physical sciences.

In reply to this criticism I would make three points. First, this criticism seems a little extreme. There is still much of population genetics which deals with the unambiguously biological: The explanation of the balanced state of sickle-cell anemia is wholly biological and manifests, beyond doubt, the hypothetico-deductive structure characteristic of the physical sciences. Moreover, I think it is true to say that population geneticists would consider their concern with polypeptides not merely an end in itself, but something which has implications for all, including the purely biological, aspects of an organism's phenotype. This certainly seems true of the balance hypothesis supporters. They cannot measure immediate phenotypic characteristics and differences, but they believe polypeptide chains tell them about them.

Second, I do not see why the polypeptide chains should not themselves be considered biological, given the population geneticists' concerns. We can treat an organism as a physical object, as when we calculate the parabola described by a human cannonball. Similarly, since the population geneticist asks not physicochemical questions of his polypeptides but biological questions—particularly, what is their adaptive value?—I suggest that it is not wrong to regard even those parts of population genetics dealing with polypeptides as biological.

Third, if it is insisted that much of population genetics is physical and not biological, then so be it. I am concerned (at present) to show that despite the complexity of biological phenomena, biology is not so very different from physics; but apparently much of my concern is otiose. Science has passed philosophy by, for areas that were previously of biological concern are now of physical concern. So why bother if biology is like physics, for physics is pushing biology aside? Who cares, as a live philosophical issue, if biology is like physics, since biology is on the road to becoming an anachronism? If nothing else, the complexity of organisms cannot be all that problematical, for although the complexity issue was raised to show that biological theories cannot be like physical theories, now apparently physical theories can go right in and explain that which biology could not! (Alternatively, if it is argued that population genetics, although a physical science, fails by normal criteria of physical science, then perhaps parts of physics are or ought to be like parts of biology after all, albeit not hypothetico-deductive. Apparently then, inasmuch as it is biological, population genetics is hypothetico-deductive, and inasmuch as it is physical, it is not.)

Of course, even if one grants everything I would claim of population genetics, this is not to deny that few other areas of biology show the same formal success or relationship between theory and evidence. As mentioned earlier, much of biology does seem very descriptive, utilizing low-grade generalizations, applying to limited classes. All I would argue here is that if anyone would maintain that this is a necessary state of affairs, he should first think carefully of the successes of population genetics and then answer why other fields are necessarily precluded from these or like successes. Biology *can* be like physics (in the respects discussed), and this is so even though not all of biology *is* like physics.

But perhaps, for two reasons, this confident conclusion is unjustified. First, as we have seen, population genetics is a statistical theory. It is clear, therefore, that nothing has been done to undermine Mayr's claim that if there are laws in biology then they must be statistical. More important, nothing has been done to show that biology contains laws and theories like the best of physical theories—nonstatistical theories like Newtonian mechanics. Second, no attention has yet been paid to evolutionary theory, the area of biology that is cited almost invariably by those who would separate biology from physics. Perhaps in that field there are special considerations of complexity and uniqueness which show that a formal hypothetico-deductive system relating throughout to the empirical evidence is impossible. It behooves me, therefore, before going on to consider my second reason for separating biology from physics, to make a few remarks about these objections.

6. NONSTATISTICAL FORMAL THEORIES IN BIOLOGY

Before turning directly to the question of nonstatistical (hypothetico-deductive) theories in biology, two points should be made. First, it must be reemphasized that the existence of such theories is not necessary to show that biology is or can be like physics. Physics contains statistical (hypothetico-deductive) theories; as we have seen, so also does biology. So the two areas of science do have some models of understanding in common, even if physics (unlike biology) has nonstatistical theories. Second, I find it far from obviously true that nonstatistical theories are in some way "better" than statistical theories. It seems to me that statistical theories capture one facet of experience, nonstatistical theories capture another facet, and there is no ultimate reason to think one facet superior to the other. It might be objected that I have allowed that presumably behind the transmission of individual genes are reasons and causes that are ignored by population genetics, thus leading to the necessity to invoke statistical understanding. Hence, it might be suggested that if one goes more deeply into the problems tackled by population genetics, a nonstatistical theory will emerge. And this would be a "better" theory than we have at present. In reply I would suggest that even if such a theory is "better" (and it certainly would not be "better" from the viewpoint of practical use) it is better because it goes into the problems in some way more deeply—it considers things ignored by the present theory—not by virtue of the fact that it is nonstatistical. In any case, one assumes that behind every or nearly every nonstatistical theory lies another theory which goes into the problems more deeply in some sense (perhaps by dealing with ever smaller elements). So the call for "better" theories is liable to rebound on all theories, statistical or not. Moreover, the nature of quantum mechanics leads one to suspect that a deeper theory may well be statistical in a way that a less deep theory is not.[34] Hence, claims about the superiority of nonstatistical over statistical theories seem most dubious.

Considering now the question of the existence or possible existence of formal nonstatistical theories in biology, if we remain with the problems considered by population genetics, then matters do not look very hopeful. Indeed, even though we might concede that nonstatistical understanding is possible if we consider organisms on an individual basis, if one wants to remain at the *group* level it is difficult to see how a statistical approach could ever be avoided. The population geneticist wants to understand the nature and *distribution* of variation in populations and the way it is transmitted from one generation to the next. The problem itself involves ratios and proportions, and thus it would seem that a statistical approach is inevitable.

But are there other areas of biology where a nonstatistical theory would seem either to exist or to be feasible? Certainly if one turns to molecular biology such a theory (or theories) seems not merely feasible but to exist: There is an elaborate theory about the way in which the information carried by the DNA can be used to make strings of amino acids (polypeptide chains), and there is nothing particularly statistical about it.[35] For example, the DNA triplet adenine-adenine-thymine (AAT) transcribes as the RNA triplet uracil-uracil-adenine (UUA), and this in turn is the code for the amino acid leucine. Although this sequence may sometimes break down, the claim that one goes from AAT to UUA and thence to leucine seems not to be statistical at all. The same seems true in the field of bioenergetics, that area of inquiry dealing with biological energy transformations. There is, for example, nothing particularly statistical about the oxidizing process of the Krebs tricarboxylic acid cycle.[36]

It might of course be objected, as was objected earlier, that molecular biological theories are really physicochemical theories, since they deal with molecules and with relations and reactions among them. In this case, molecular biological theories hardly tell us much about the similarities and differences between physical and biological theories (except inasmuch as they tell us about physical theories!). I am no happier with this line of argument now than I was previously. If one considers problems about the way in which the information encoded in DNA is passed from one generation of organisms to the next, one is tackling problems that seem to have as much right as any to be called "biological." But even if one allows the objection, the same kind of comment is appropriate here as was appropriate previously. In particular, one cannot get away from the fact that molecular biological theories, even if they are physicochemical, exist and apply to biological phenomena.

In itself there is nothing particularly surprising about this fact, for other physicochemical theories apply to biological phenomena—falling elephants obey Galileo's laws of motion just as do falling rocks.[37] But it does mean that the extent to which biological objects can be treated as physicochemical objects has increased dramatically in recent years and is still increasing. Moreover, a physicochemical theory like molecular genetics differs from a physicochemical theory like Galilean mechanics in this respect: Whereas the application of Galilean laws of motion to biological phenomena does not touch biological theories in any significant way, a theory like molecular genetics does seem to displace biological theories, specifically Mendelian genetics, in some way. As it were, biological problems about biological properties (for example, inheritance and gene action) are starting to give way in importance to physicochemical problems about molecular properties (for example, DNA action and inheritance). Hence, if nothing else, the question of the autonomy of biology suddenly no longer seems so very important, for biology as a live progressive discipline seems to be on the decline.[38]

If we leave the rather controversial area of molecular biology and turn to the undeniably biological, is there no area at all that offers hope of a formalized axiomatic theory incorporating nonstatistical laws? One area that at least looks inviting is that of organic ontogenetic development (the development of individuals rather than groups), and there is certainly some promising work in that field. In particular, Aristid Lindenmayer and his associates are trying to develop formal systems which will capture essential aspects of organic development at the level of the cell.[39] Working with ideas and techniques very similar to those used in abstract language theory, they have been able to produce systematically related claims about functions with known mathematical properties, functions which can be used to generate models closely analogous to the early developmental stages of both some plants and some animals. Thus, for example, one function can be used to simulate the growth of a leaf with lobes (figure 2.a).[40] This function has some rather interesting properties. For instance, it shows that although the final finished leaf may be symmetrical, its growth will be asymmetrical, something which is actually found to happen in nature. It shows also that were the central growing point of the leaf to die at some intermediate stage, the initiated lobes would continue to expand. This is again something in fact found to happen (figure 2.b).

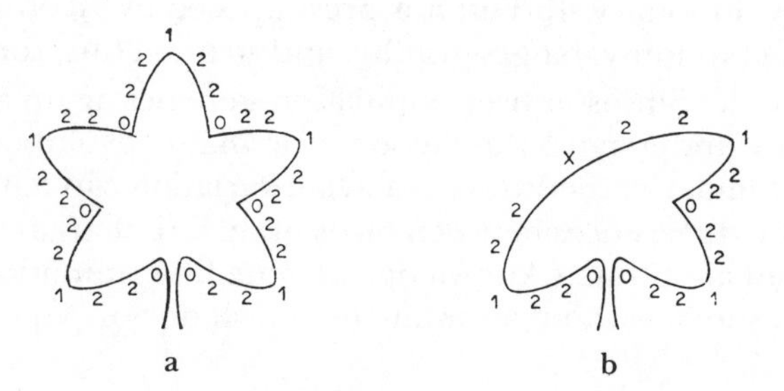

Figure 2. Outlines of Fully Developed Leaf and of Leaf with Dead Growing Point

It would, I think, be misleading to pretend that we have here a fully articulated formal system with wide experimental evidence (there is, for example, no question of showing how a zygote turns into a man), but it does seem to be a rapidly expanding area in which formal techniques are being used successfully to bring understanding to biological phenomena. Moreover, there is nothing inherently statistical about this theory. Although it can be extended to cover stochastic processes where statistical methods may be required, it can be restricted and applied where no such processes are involved. In short, this area of biology may well in the future give the lie to Mayr's claims that even if there are genuine laws in biology, they must be statistical. Perhaps here we do indeed have the makings of a nonstatistical, biological, hypothetico-deductive theory.

7. EVOLUTIONARY THEORY

I come now to the much debated question of the nature of evolutionary theory. Many philosophers (and some scientists too) argue that evolutionary theory is the paradigm case of a theory which is not, and in principle never could be, a theory like those to be found in physics.[41] It is argued that the complexity and uniqueness of evolutionary phenomena preclude physical-science-type understanding, and many other "distinctively biological" models have been offered instead (I mentioned three such models earlier). These models all try to weaken physical-type explanation in some way in order to cope with what are thought to be peculiarities of the evolutionary scene.

I shall not here address myself directly to these models; I have already done this elsewhere.[42] What I want to do chiefly at this point is to suggest that much use that philosophers make of evolutionary theory, including that of putting it against the physical sciences, is misguided. It stems, I believe, from a fundamental ignorance of the true nature of evolutionary theory. Many philosophers read the *Origin of Species* and leave matters at that. They are entirely ignorant of the importance for modern evolutionary studies of the development of genetics, particularly population genetics. What I would argue is that today population genetics lies at the heart of evolutionary studies; it is presupposed by all other areas like paleontology, taxonomy, biogeography, and so on. Thus, for example, a paleontologist like Simpson uses population genetics again and again as he attempts to understand the reasons for major evolutionary trends and changes. Similarly, the student of avian variation can hardly begin to understand the differences between birds of different areas (like oceanic islands) without an intimate knowledge of gene flow, mutation, selection, and so on—in short, without an intimate knowledge of population genetics.[43]

But if this characterization of evolutionary theory which I have just offered is accepted, then much of the sting from the charge of separateness vanishes. Inasmuch as population genetics participates in the characteristics of a physical science, so also does evolutionary theory. Admittedly much of evolutionary theory is diffuse and rather descriptive, but a part—the essential part—is not.

The listener might be feeling that I am making rather heavy weather of all this, particularly since I appear to be ignoring a recent, heaven-sent opportunity to argue that evolutionary theory is as formally hypothetico-deductive as any theory of the physical sciences. I refer to Mary Williams's axiomatization of evolutionary theory.[44] Williams starts with a number of explicitly stated axioms and then, as rigorously as you like, merrily deduces consequences about evolutionary happenings. Surely, here again science has overtaken philosophy. My arguments are unneeded, for evolutionary theory now has all the marks of a good physical theory.

The catch to this line of reasoning, of course, is whether Williams's axiomatic theory really is evolutionary theory or a satisfactory substitute. As anyone who has worked through any of J. H. Woodger's laborious calculations will know well, axiomatization, be it ever so rigorous, is worth little if it is done only at the expense of most of the scientific content.[45] On the other hand, Woodger's basic problem is that he encumbered himself with the crudest of crude empiricisms, and this is certainly not a fault with which one can charge Williams. So it would be unfair to tar her with the same brush as one would Woodger, solely because they are both axiomatizers. Nevertheless, even after one has made such an effort to be fair, I am still far from convinced that Williams's axiomatization is a panacea for all of evolutionary theory's woes, real and imaginary.

My main unease with her axiomatization stems from its treatment, or rather its nontreatment, of the problems of heredity. Even if one does not give population genetics quite the central role in evolutionary studies that I do, it is still clear that a major concern of the evolutionist will be heredity. What causes phenotypic difference? Where does it come from? How does it get passed on? Without some answers to some questions like these, the evolutionist is in a hopeless mess as he tries to explain the change in organisms over periods of time. And if anyone doubts the truth of this, he should recall Darwin's desperate and somewhat pathetic struggles to supplement his theory of the *Origin of Species* with a theory of heredity.[46]

Now, of course, to fill the gap facing Darwin, we have Mendelian genetics and its modern counterpart, molecular genetics. However, Williams in her axiomatization avoids nearly all mention of the problems of heredity. Without going into any great detail, the way in which her axiomatization runs is as follows. She begins by introducing the notion of a "biological entity." This is something subject to certain restrictions, for example, it cannot be a parent of itself, but it is also something which she deliberately keeps at a high level of generality. It can apply indifferently to a gene, an organism, or a group of organisms—anything in fact which satisfies her restrictions. Williams then introduces her main axioms. These axioms bring in the notions of limited potential for population growth, of organic fitness, of selection, and so on. Now it is certainly the case that her axioms make stipulations and restrictions that any theory of heredity (to be used in conjunction with her theory) would have to satisfy. For example one axiom (5) specifies that under certain conditions subgroups within groups will increase relative to the whole group, and this would seem to imply that certain peculiar variations can be preserved and transmitted to future generations. But, as she herself openly admits (and indeed looks upon as a virtue of her theory), Williams does not make any commitment to any particular theory of heredity. In particular, she does not commit herself to the modern molecular-cum-Mendelian theory. And indeed, it is difficult to see how, within her theory, she could do so, without altering it radically. We have just seen that Williams keeps her discussion very general, wanting to feel free to interpret the basic unity,the biological entity, as anything from a gene, up and through an organism, to a group of organisms like a species. Were she to spell out things specifically in terms of units of heredity, she would lose this generality.

It is therefore necessary to bring to Williams's axiomatization our knowledge of genetics. But what happens when we do? Suppose, using this knowledge, we interpret her axioms in terms of Mendelian genes and the relationships that hold between such genes. We can then derive theorems about these genes, including, as she proudly points out, a theorem that given within a population a pair of alleles with heterosis (that is, heterozygote fitter than either homozygote), then both alleles will be maintained in the population in a balance. However, is it unfair to ask why this should excite us, because population geneticists know, and can and do deduce this fact already? And this question brings me to the nub of my criticism of Williams's axiomatization. We have to bring to it our knowledge of genetics; but if once we do so, I see no evidence that it is any more powerful to crank out theorems than present techniques of population genetics. We have already achieved axiomatization at this level; Williams seems at best to be duplicating it.

But perhaps, despite my earlier promise, I am being unfair. Perhaps Williams is not really trying to give us a surreptitious substitute for population genetics. Let us therefore interpret her axioms in perhaps the most obvious way, as being about organisms subject to certain basic rules of inheritance, and then accept her theory for what it is, namely, a system which shows how over a period of time one group of organisms can be replaced by another, and so on.[47] This her theory certainly shows, and if this is all that she would claim of her theory, then credit is certainly due to her for showing this much. But let us not pretend that she has offered us an immediate answer to all the evolutionist's problems, a way in which all of his thought can at once be put in a satisfactory hypothetico-deductive mold.

Take, for example, the kind of problem that might face the paleontologist, say the explanation of the evolution of the horse. One certainly cannot just pop the eohippus in at one end of Williams's theory and drag out the modern horse at the other end. Not surprisingly, Williams's theory demands that one assign certain fitness coefficients to evolving organisms. Unless one does this, there is just no way in which one can calculate which organisms will evolve, which will remain static, and which will become extinct without successors. But the trouble is no one knows that much about the kinds of selective forces facing the eohippus and his ancestors, even though some plausible guesses can be made about some of them. Certainly there is no way in which we can quantify any of these forces—as we have seen, it is bad enough having to try to pin down the selective forces working on today's fruit flies. Hence, if nothing else, one runs into immediate problems with trying to apply Williams's theory to real-life situations.

At this point Williams would probably gripe that my complaint seems to be about axiomatic evolutionary theories generally, not about hers in particular, and in this I think it would be true to say she is correct. It just seems to me that the massive problems facing the evolutionist—vast periods of time, incredibly complex situations, so much vital information irretrievably lost—preclude solution by quick axiomatization. Either the axiomatizations are inapplicable, or they apply so insensitively that they tell us little. To some relatively simple situations evolutionists already apply axiomatic techniques, and no doubt they will continue to extend these in the future. But satisfactory full-blown evolutionary axiomatizations seem a faint hope, now or in the immediate future.

I have dealt somewhat unsympathetically with Williams's axiomatization. I do not want to leave the impression that what she has done is of absolutely no value. One point of great interest is that because of the generality of reference of her axioms and theorems, as she herself points out, her theory may well indicate valuable but unsuspected analogies between different levels of biological activity—for example, between gene balances and predator-prey balances. What I would argue here is that one should not seize on her theory as definitive proof that evolutionary biology is like physics.

8. TELEOLOGY

I come now to the second reason purportedly separating biology from the physical sciences—that there is in biology, unlike the physical sciences, an irreducible, irremovable *teleological* element. In biology, so the argument goes, it makes sense to ask what "function" something has, and to say of something that it exists or performs "in order that" something else may happen. Such a way of thinking or speaking is alien to the physical sciences. One cannot, or at least ought not, ask what function the moon has; and the moon does not exist in order to make the tides work (nor does it exist for any other purpose). Hence, considered from the viewpoint of teleology, biology is different from physics.

Let me state at once that I do believe biology is to be distinguished from physics in that it, unlike physics, has an irreducible teleological element.[48] I think that in biology one has an understanding through ends or effects, a kind of understanding that one does not have in physics. More specifically, I would suggest that when a biologist uses teleological language he is referring in some way to an adaptation, that is to say, to something which helps its possessor in the struggle to survive and reproduce, and the reason the language is teleological is that he is interested in what way the adaptation helps its possessor, and because the biologist uses this way to understand the adaptation. Thus, when the biologist states that the function of the heart is to circulate the blood, he is showing how the heart, the adaptation, helps in survival and reproduction, namely through blood-pumping. Similarly, when the biologist states that female breasts exist in order to suckle the young he is explicating the way in which adaptations, the breasts, help survival and reproduction. Now what seems to me to be distinctive here is that we try to analyze or understand one thing, the adaptation, in terms of its ends or effects. We do not say the circulation of blood or the suckling of young exist for the benefit of the heart or breasts, but vice-versa. In other words, we are trying to explain or understand the *cause,* the heart or breasts, in terms of the *effects,* the circulation of blood or the suckling of young.

This kind of explanation or understanding does not occur (or at least occurs only in a very limited fashion) in the physical sciences.[49] We do not explain the cannonball's hitting the wall in terms of the subsequent bang it causes, but rather we explain the bang, the effect, in terms of the impact, the cause. Hence, for this reason it does seem to me that there is something distinctively different about a certain kind of biological understanding. Furthermore, although one certainly may be able to avoid function-talk by, for instance, always talking directly in terms of adaptation and adaptive advantage, I do not really think that one can translate the teleological element right out of biology without losing something. This is not to deny that, even in biology, one can always and only look at effects from the viewpoint of their causes. Thus, for example, one might explain present breasts, the effects, in terms of past reproductive successes of ancestors with breasts which could suckle young, the causes.[50] But it does seem to me that this is not the same thing as looking at causes from the viewpoint of their effects; and as I understand it, translation does involve saying or having the "same thing." Certainly, when I say that my heart exists in order to circulate my blood, although it is indeed true that my heart is in a sense the effect of my ancestors' circulating blood, it does not seem that what I am doing is trying to relate my heart to my ancestors' blood (my heart does not exist to circulate their blood), but rather I am trying to relate my heart to *my* blood. I am after different kinds of understanding in the teleological and nonteleological cases.

A number of points arise from the thesis about biological teleological understanding that I am presenting here, and so I shall look briefly at a number of them. First, since teleology is usually thought to have something to do with the future, it is worth pointing out in what sense I am suggesting that the biologist is concerned with, more specifically attempting to explain through, the future. One thing I am certainly not suggesting is that the biologist is committed to future causes, in the sense of things acting backward through time. Although we might loosely ask "What is the cause of breasts?" expecting an answer like "the suckling of young," it is the breasts that are the cause, not the suckling. (To put the matter in traditional language, the sense of "cause" in "what is the cause of breasts" is that of "final cause" not "efficient cause." The biologist explains in terms of final causes, not in terms of backward-working efficient causes.) However, I do agree that the biologist is trying to understand through the future (unlike the physicist). I understand the present breasts through their future effects, the suckling.

Two qualifications should be added. When I speak of "future," I do not necessarily mean our future. I mean future relative to the adaptation, the cause, which is said to have the function. Thus, the paleontologist may be trying to understand causes relative to their future effects, but effects which are long past to us. Second, if one grants that causes and effects can occur simultaneously, perhaps a pumping heart and circulating blood, then one has little future reference. But one still has the understanding of causes in terms of effects which I believe is distinctive of biological teleology.

A second question that might be asked is how it comes about that biology, unlike the physical sciences, has this teleological element? From a historical viewpoint this is easy enough to answer. Darwin took it over directly from his nonevolutionary predecessors, and here it has stayed.[51] But how did it come to be in pre-Darwinian biology in the first place? To answer this question it is necessary to recognize the grip that the argument from design held on the pre-Darwinian imagination.[52] Everything was seen as the product of a good god's creative endeavor—the heart, the hand, and above all, the eye. But since all was created by God, it makes sense to use teleological language and thought, for teleological language and thought are what we use when conscious intelligences have been at work. Artifacts are made with some purposes or ends in mind, and consequently we understand them functionally, that is, with respect to these ends. Thus, a pump is designed to drive water up a pipe, and certain valves have the function or end of preventing the water from slipping back, and we understand the pump and valves in terms of these ends. In a similar manner, since organisms and their characteristics are God's artifacts, functional language is not merely in order but demanded. The eye was literally intended for sight; thus, functional language is appropriate for the eye.

A couple of questions arise. First, if functional language was appropriate for organisms, how was it that pre-Darwinians, who certainly believed that God designed the inorganic world as well, did not use function-talk in their physics? Two comments may be made. First, although by the nineteenth century teleological language had gone from formal physics, even some physicists were quite happy to view the physical world teleologically. Thus, for example, Sir David Brewster, biographer of Newton and inventor of the kaleidoscope, argued that the function (or at least, a function) of the sun and the moon is to light the earth for its (human) inhabitants.[53] Second, the reason there was a difference *qua* function-talk in the formal sciences of physics and biology is that people felt that the evidence of design is far more overwhelming in the organic world than it is in the inorganic world. Design could be found in the inorganic world, but one had to argue carefully to it. One did not have to bring it into one's physics, and indeed there was no overriding compulsion to do so. In the organic world, on the other hand, it just thrust itself at you—"irrefragible evidence of creative forethought," as the anatomist Richard Owen said of the adaptations of the kangaroo for feeding its young—and thus one just could not avoid talk of design.[54]

This schizophrenic attitude toward design is brought out most clearly in the thought of William Whewell. Whewell was so taken with organic evidences of design that he argued we must presuppose it even to have biological understanding.

> Thus we necessarily include, in our Idea of Organization, the notion of an end, a purpose, a design; or, to use another phrase which has been peculiarly appropriated in this case, a *Final Cause.* This idea of a Final Cause is an essential condition in order to the pursuing our researches respecting organized bodies.
>
> This Idea of Final Cause is not deduced from the phenomena by reasoning, but is assumed as the only condition under which we can reason on such subjects at all.[55]

But when it came to the inorganic world, Whewell played a very different tune: "The impression of benevolent design in [the inorganic] case is less striking and pointed than that which results from the examination of some other parts of nature."[56] (These other parts are the organic parts.) Consequently, Whewell wrote:

> That there is a purpose in many other parts of the creation, we find abundant reason to believe from the arrangements and laws which prevail around us. But this persuasion is not to be allowed to regulate and direct our reasonings with regard to inorganic matter, of which conception the relation of means and end forms no essential part. In mere Physics, Final Causes, as Bacon has observed, are not to be admitted as a principle of reasoning.[57]

The second question arising from the role of the argument from design in pre-Darwinian biology is, granting that the pre-Darwinians were teleologists because they accepted the design argument, and granting that Darwin got his teleology from them, why should it have been possible for Darwin to put the teleology right into his theory? Surely one did not have to accept the argument from design to be a good Darwinian? The whole point of what Darwin showed was that one did *not* have to accept immediate creative design for organisms; at best, God designed organisms through the natural agency of natural selection working on random mutation, and given this natural mechanism there is no obligation to accept God's design at all.

The reason I think Darwin could use design-talk (and the reason modern biologists continue to do so) is that although organisms may be the product of natural selection on random mutation, the effect is like objects of design—man's design, that is. The eye *is* like a telescope, and the heart *is* like a pump, even though God may have designed neither. Thus, Darwin and modern biologists could and can continue to exploit the analogy or metaphor of design. They find it profitable to treat the organic world as though it were a collection of artifacts, even though they may not believe organisms really to be artifacts, not even God's artifacts.

That my claim is borne out historically is, I think, shown by Darwin's argument in the *Origin.* He likens the effect of natural selection in wild organisms to the effect of man's selection in domestic organisms, and time and again we are invited to see wild organic ends in the light of ends that man might have.[58] Thus, the peacock's feathers have the end of beauty, of attracting the peahen, just as the magnificent feathers on man's domestic productions are to be understood in terms of man's aesthetic enjoyment. In other words, Darwin tries to see the peacock in terms of human design, of domestic variations (and if anyone does not think domestic variations are products of design, they should read the grandiose claims, real and apparent, that breeders in Darwin's time were making.)[59] More generally it is worth adding that Darwin admitted openly to having been much influenced by Archdeacon Paley on design, and thus Darwin had long been accustomed to thinking of organic characteristics as if they had been consciously formed for some use.

I think also that modern biologists look upon organisms as if they are artifacts, and hence they feel free to use teleological language. Thus, for instance, in a much praised book, G. C. Williams writes:

> Whenever I believe that an effect is produced as the function of an adaptation perfected by natural selection to serve that function, I will use terms appropriate to human artifice and conscious design. The designation of something as the *means* or *mechanism* for a certain *goal* or *function* or *purpose* will imply that the machinery involved was fashioned by selection for the goal attributed to it.[60]

And then Williams explicitly exhorts biologists to try to solve their functional problems by looking at organisms as if they were (human) artifacts. Finally, the whole metaphor of the "genetic code" would seem to imply that the artifact analogy or metaphor has found its way into the molecular biological world also.

I suggest, in short, that biological function-talk is appropriate because organisms look like objects of conscious design. It is felt inappropriate in physics because inorganic objects do not look like objects of conscious design.[61]

9. THE PROBLEM OF GOAL-FAILURE

Let us now grant that biological teleology is genuine teleology. The final, perhaps most important, question to be asked is why biologists are not plagued by a problem that is thought to haunt teleological analyses. This problem comes under such guises as the problem of the "missing goal-object" or "goal failure," a good paradigm for which is yielded by the cat waiting outside the mouse-hole.[62] If the mouse appears, then teleologically we explain the cat's waiting in terms of the poor mouse's fate. But if no mouse appears, then the mouse's unfortunate fate does not exist; hence it can hardly explain the cat's waiting. Thus, we seem forced to explain the identical waiting cat in two different ways, something which at the very least is thoroughly counterintuitive, if not outrightly absurd. Obviously this kind of thing never occurs in normal explanations, because by the time we are explaining an effect, the cause must have occurred; it cannot be missing or have failed!

Let us see how goal failure might arise in the functional case. Take the male genitalia as a paradigm instance of things with a function, namely their role in sexual intercourse leading to reproduction. What about cases of goal failure? (I hope any Catholic priests will forgive my taking their celibacy as a paradigm case of goal failure. I speak biologically, not spiritually.) In one sense I suppose one would have to say that anyone who tried to explain the celibate's genitals in terms of intercourse and reproduction was mistaken: It is hardly the case that they existed in order to be used to reproduce. And yet, on the other hand, this kind of goal failure does not seem very troublesome to the biologist. The reason is surely that when the biologist talks of function and purpose, unless he specifically states otherwise, it is the *type,* not the *individual,* that he has in mind.[63] "The purpose or function of the male genitalia is to help their possessors reproduce"[64] is thought true, not because every possessor of male genitalia has intercourse and reproduces, but because a significant fraction do. What would be "significant" here might vary—it could be a very, very small percentage in the case of some functions. What would be ruled out would be the isolated freak case where, because of strange circumstances, some abnormality fortuitously proved useful. Function-talk is appropriate only where selection takes up and preserves a characteristic, be it a characteristic of ever-so-small a percentage of the population.

I suggest that this concern with the type rather than the individual makes the goal-failure problem relatively harmless for the biologist. Failure in the individual case does not touch his analysis; only a general failure affects him. Of course a general failure is not logically impossible. A drastic change in environmental conditions like the introduction of a new competitor might mean that an adaptation never again has the functional effect ascribed to it. But such a failure is unlikely to be so rapid as to catch the biologist completely flat-footed. Moreover, remember that the biologist often—always in the case of the paleontologist—will be working with organisms where both adaptational causes and functional effects are in the past with respect to his analysis. Thus, for example, the paleontologist might explain the horns of the Irish elk in terms of their value when the animal is in rut, and he would be quite secure from cases of future goal failure, because the rutting periods of Irish elks are long past, as also are the horns which he is trying to explain.

One might ask what the biologist would do if he were faced with a case of goal failure. A characteristic had a function but, perhaps because of changed environmental conditions, no longer has a function. In a case like this I think it is obvious that the biologist would explain the characteristic nonteleologically, in terms of *past* selective processes. He would argue that the organism has the characteristic because the organism's ancestors found it selectively advantageous to have the characteristic.

Since this nonteleological analysis would seem to be available in just about every case where an ascription of function is involved, it might be argued that it is really this nonteleological analysis that the biologist is offering whenever he offers a functional analysis. Thus, so this argument would run, when the biologist explains a characteristic x in terms of its function y, it is not the future effect of x which is involved, but all the past y's of the characteristic's possessor's ancestors. In other words, the teleology of the biologist is really quasi teleology, because what he intends, despite his language, is a normal causal explanation.[65]

Although I agree that the biologist can usually offer this nonteleological analysis as a substitute for his functional analysis, and that hence one could eliminate the teleology of biology, I would disagree that the biologist normally intends this substitution, or, consequently, that the biologist's teleology is really quasi teleology. To a great extent I cannot really argue my position. I just think the way I have described matters is the way they really are. However, there is a kind of borderline situation that supports what I would claim. Suppose a character, unadaptive previously, suddenly becomes adaptive (or becomes adaptive in a new way). The kind of situation I have in mind is where an organism or a handful of organisms breaks into a new environment (for example, birds blown by a storm to an island). In this situation the biologist would say that character *x* has function *y*, and *x* would be explained in terms of *y*, although *x* would certainly not exist because the organism's ancestors had *x*'s which were adaptive because of their *y*-performance. In other words, in this case the functional analysis could not be replaced by a nonteleological causal analysis.[66] Hence, since it is clear that some functional analyses are *not* considered equivalent to, or shorthand for, nonteleological analyses, I see no reason to assume that they ever are—even though in most cases a functional analysis could certainly be replaced by a nonteleological analysis.

Obviously this claim I am making presupposes that biologists really are prepared to talk of "function" as soon as a character starts to prove adaptive. I cannot speak for all biologists, but the following passage by G. G. Simpson shows that at least some biologists hold the position I presuppose:

> In order to realize the new functions of a changed environment, an organism must, at the moment when the change or the occupation of a new environment begins, have at least some functions prospective with regard to the new environmental functions. This is merely a more technical way of saying that organisms can continue to live only under conditions to which they are already at least minimally adapted.

And then he adds later:

> If a mutation, whether adaptive, non-adaptive, or inadaptive with respect to the adaptation of an ancestral population does become fixed and spread by selection it is adaptive *from the start* with respect to the descending populations.[67]

10. CONCLUSION

I have reached the end of my essay. In answer to the question I posed in my title, I reply that yes, in some respects biology is different from physics—much of it is more loosely organized and it has an irreducible teleological element; but no, in other respects biology is not different from physics—biologists, particularly population geneticists, want to explain using laws in a hypothetico-deductive manner, and they have at least some success; biologists make no use of fancy kinds of backward-working efficient causes, and indeed, if it were so desired it does seem that one could eliminate function-talk from biology. In short, all other things being equal, biology is like physics—but not quite everything is equal.

NOTES

1. In my book *The Philosophy of Biology* (London: Hutchinson, 1973), I consider other reasons.

2. Variants of this argument occur in T. A. Goudge, *The Ascent of Life* (Toronto: University of Toronto Press, 1961); G. G. Simpson, *This View of Life* (New York: Harcourt, Brace and World, 1963); W. M. Elsasser, *Atom and Organism* (Princeton: Princeton University Press, 1966); B. Glass, "The Relation of the Physical Sciences to Biology—Indeterminacy and Causality," in *Philosophy of Science,* The Delaware Seminar, vol. 1, ed. B. Baumrin (New York: Interscience, 1963), pp. 223–49; and E. Mayr, "Cause and Effect in Biology," *Science,* 134 (1961):1501–06. A good critical introduction to this problem is provided by K. F. Schaffner, "Antireductionism and Molecular Biology," *Science,* 157 (1967):644–47.

3. I adopt here without discussion the standard "logical empiricist" analysis of physical theories (see R. B. Braithwaite, *Scientific Explanation* [London: Cambridge University Press, 1953]; E. Nagel, *The Structure of Science* [London: Routledge and Kegan Paul, 1961]; and C. G. Hempel, "Aspects of Scientific Explanation," in his *Aspects of Scientific Explanation and Other Essays in the Philosophy of Science* [New York: The Free Press, 1965]). I am aware of the arguments of critics such as P. Achinstein (*Concepts of Science* [Baltimore: Johns Hopkins Press, 1968]; *Law and Explanation: An Essay in the Philosophy of Science* [London: Oxford University Press, 1971]), but they seem to me not to touch the essential claims of the logical empiricists. In any case, many of the critics (for example, Goudge, *Ascent of Life*) use biology to fight logical empiricism, and these I try to counter in this essay—insofar as I want to counter them.

4. Strictly speaking, something very close to Kepler's laws is inferable, and the deduction also requires mathematical premises.

5. J. D. Watson, *Molecular Biology of the Gene,* 2d ed. (New York: Benjamin, 1970), p. 85.

6. The first two models are suggested in Goudge, *Ascent of Life,* and the third in W. B. Gallie, "Explanations in History and the Genetic Sciences," *Mind,* 64 (1955):160–80.

7. Indeed, I shall argue later that it might be in the interest of the defender of biological peculiarity to separate these two arguments.

8. Mayr, "Cause and Effect," p. 1505. Surprisingly, Mayr does allow the existence of biological statistical laws. If he really means what he claims about uniqueness, then these could not exist. Given a statistical law, "x percent of A's are B's," we know at least that x percent of A's are nonunique, inasmuch as they are all also B's.

9. G. G. Simpson, "Historical Science," in *The Fabric of Geology,* ed. C. C. Albritton, Jr. (Stanford, Calif.: Freeman, Cooper and Co., 1963), p. 30.

10. W. M. Elsasser, "Quanta and the Concept of Organismic Law," *Journal of Theoretical Biology,* 1 (1961):27–58.

11. For example, Schaffner, "Antireductionism"; N. Rescher, *Essays in Philosophical Analysis* (Pittsburgh: University of Pittsburgh Press, 1969). A more sympathetic treatment of the problem of complexity can be found in W. Wimsatt, "Complexity and Organization," in *PSA-1972* (selected proceedings of the 1972 Philosophy of Science Association), Boston Studies in the Philosophy of Science, ed. K. F. Schaffner and R. S. Cohen (Dordrecht: Reidel, 1973).

12. Simpson, "Historical Science."

13. Alleles are variant genes that can occupy the same locus on a chromosome.

14. A good recent introduction to population genetical theory is J. F. Crow and M. Kimura, *An Introduction to Population Genetics Theory* (New York: Harper and Row, 1970). I myself discuss some of the elementary parts of the theory in "Are There Laws in Biology?", *Australian Journal of Philosophy,* 48 (1970):234–46; and *The Philosophy of Biology.* The reader will have to check for himself the general truth of my claim that the inferences in population genetics are deductive, but one simple example may help to put the not-too-dubious reader at ease.

Suppose we have complete selection against a recessive allele, that is, suppose we have an allele A_1 which shows itself in the phenotype only when the genotype is $A_1 A_1$, and that this phenotype always fails to reproduce. Suppose moreover that we have a large, randomly mating group (that is, one to which the Hardy-Weinberg law can be applied) and that the initial proportion of A_1's is P_o. It follows (by the Hardy-Weinberg law) that there will be P_o^2 $A_1 A_1$'s—all of which will fail to reproduce. Hence, in the next generation, the only $A_1 A_1$'s will come from the mating of *two* organisms carrying one A_1 allele recessively—¼ of the offspring (by the Hardy-Weinberg law) will be $A_1 A_1$. Since (by the Hardy-Weinberg law) among organisms which will breed, the fraction of organisms carrying one and only one A_1 is

$$2P_0(1-P_0)/((1-P_0)^2 + 2P_0(1-P_0)),$$

the proportion of $A_1 A_1$'s in the next generation is

$$\tfrac{1}{4}\,([2P_o(1-P_o)]/[(1-P_o)^2 + 2P_o(1-P_o)])^2 = [P_o/(1+P_o)]^2.$$

Hence the proportion P_1 of recessive genes in this next generation is

$$P_0/(1 + P_0).$$

More generally, if the proportion in the ith generation is P_i, then the proportion in the i + 1th generation is given by the formulae

$$P_{i+1} = P_i/(1+P_i) = P_o/[1+(i+1)P_o].$$

Clearly, the reasoning to this conclusion has been deductive.

15. A brief exposition of the kinetic theory of gases, together with the details of this example, can be found in A. Ingard and W. L. Krasuhaar, *Introduction to Mechanics, Matter, and Waves* (Reading, Mass.: Addison Wesley, 1960).

16. The theory of balance due to heterosis and details of this example are given in my *Philosophy of Biology.*

17. E. Nagel, *The Structure of Science* (London: Routledge and Kegan Paul, 1961), p. 281.

18. The basic theory of genetic drift was worked out by Sewall Wright, "Evolution in Mendelian Populations," *Genetics,* 16 (1931):97–159, and it is indeed sometimes called the "Sewall Wright effect." Crow and Kimura, *Introduction to Population Genetics Theory,* give modern theoretical treatments of drift. There seem to be analogies here with the physical theory dealing with Brownian motion. In particular, one is reminded of random walk with absorbing barriers (S. Chandrasekhar, "Stochastic Problems in Physics and Astronomy," *Review of Modern Physics,* 15 [1943]:1–89).

19. J. J. C. Smart, *Philosophy and Scientific Realism* (London: Routledge and Kegan Paul, 1963).

20. Ruse, "Are There Laws in Biology?"; idem, "Is the Theory of Evolution Different? I The Central Core of the Theory. II The Structure of the Entire Theory," *Scientia,* 106 (1972):765–83, 1069–93; and idem, *The Philosophy of Biology.*

21. R. A. Fisher, "Has Mendel's Work Been Rediscovered?", *Annals of Science,* 1 (1936):115–37; S. Wright, "Mendel's Ratios," in *The Origins of Genetics: A Mendel Source Book,* ed. C. Stern and E. R. Sherwood (San Francisco: W. H. Freeman, 1966), pp. 173–75.

22. R. C. Lewontin, *The Genetic Basis of Evolutionary Change* (New York: Columbia University Press, 1974), is an excellent introduction to this controversy, and a good reference source. Where no other references are given, the reader should assume that I refer to Lewontin.

23. See H. J. Muller, "The Darwinian and Modern Conceptions of Natural Selection," *Proceedings of the American Philosophical Society,* 93 (1949):459–70; "Our Load of Mutations," *American Journal of Human Genetics,* 2 (1970):111–76. As I shall point out shortly, today's followers of Muller advocate a somewhat different hypothesis than his "classical" hypothesis.

24. See T. Dobzhansky, *Genetics and the Origin of Species,* 3d ed. (New York: Columbia University Press, 1951); idem, *Genetics of the Evolutionary Process* (New York: Columbia University Press, 1970); and E. Mayr, *Animal Species and Evolution* (Cambridge, Mass.: Belknap, 1963).

25. See A. B. Raper, "Sickling and Malaria," *Transactions of the Royal Society of Tropical Medicine and Hygiene,* 54 (1960):503–04; F. B. Livingstone, *Abnormal Hemoglobins in Human Populations* (Chicago: Aldine, 1967); "Malaria and Human Polymorphisms," *Annual Review of Genetics,* 5 (1971):33–64; M. Ruse, "Some Thoughts on Programmes for Improving Mankind," *Proceedings of the Fifth Conference on Value Inquiry,* forthcoming; idem, "Reduction in Genetics," in *PSA-1974,* Boston Studies in the Philosophy of Science, ed. A. C. Michalos and R. Cohen (Dordrecht, Holland: Reidel, 1976):633–51.

26. Lewontin, *Genetic Basis,* pp. 89, 92.

27. Ibid., p. 113.

28. As was seen, genetic drift is gene change due to chance, that is, to random sampling. M. Kimura and T. Ohta (*Theoretical Aspects of Population Genetics* [Princeton: Princeton University Press, 1971]) give the most detailed analysis and defense of the neoclassical position.

29. B. Spassky, T. Dobzhansky, and W. W. Anderson, "Genetics of Natural Populations. XXXVI. Epistatic Interaction of the Components of the Genetic Load in Drosophila Pseudoobscura," *Genetics,* 52 (1965):653–64. I must add, particularly since I shall shortly indicate a personal preference for the balance position, that the classical-hypothesis supporters have other arguments against the balance hypothesis. One argument is based on the supposed fact that all organisms have evolved molecularly, at approximately the same rate. This, they suggest, can have been due only to drift.

30. F. J. Ayala, M. L. Tracey, L. G. Barr, J. F. McDonald, and S. Pevez-Salas, "Genetic Variation in Natural Populations of Five Drosophila Species and the Hypothesis of the Selective Neutrality of Protein Polymorphisms," *Genetics,* 77 (1974):343–84; p. 379.

31. Ibid., p. 381.

32. Between the writing of the first and second drafts of this paper, Tracey and Ayala have produced evidence suggesting that inbreeding depression is not necessarily a problem for the balance hypothesis. In particular, they suggest that the fitness of flies *Drosophila melanogaster,* homozygous for one chromosome, is compatible with the hypothesis that variability at more than 1,000 loci is maintained through heterotic selection (M. C. Tracey and F. J. Ayala, "Genetic Load in Natural Populations: Is It Compatible with the Hypothesis that Many Polymorphisms Are Maintained by Natural Selection?", *Genetics,* 77 [1974]:569–89).

33. This is not to deny that extrascientific considerations intrude. For instance, balance-hypothesis supporters are not above implying that classical-hypothesis supporters are racists for thinking in terms of an ideal type, and the term "Platonist" is gaily hurled by all. It is indeed a paradox that the name of Plato, the greatest of all philosophers, should form the basis of the greatest insult one biologist can give another.

34. Nagel, *Structure of Science.*

35. Watson, *Gene.*

36. A. Lenninger, *Bioenergetics* (New York: Benjamin, 1965).

37. For this reason, those who argue for the autonomy of biology on grounds of biological complexity and uniqueness ought carefully to point out that they are not, or at least ought not to be, arguing that in every respect every biological phenomenon is complex and unique. There is nothing very complex or unique about the motion of a falling elephant. What, if anything, is unique and complex about biological phenomena stems from their *biological* properties and relations.

38. Paradoxically, since molecular genes (strips of DNA) can be distinguished by different base orders, one might have, where different base orders code for the same amino acids, different molecular genes but Mendelian genes which have to be regarded as identical. Hence, if anything, Mendelian genetics seems less attuned to the complex and unique than does molecular genetics. In other words, if molecular genetics is physicochemical but Mendelian genetics is biological, physics is *more* sensitive to the complex and the unique in organisms than is biology!

39. A. Lindenmayer, "Mathematical Models for Cellular Interactions in Development," *Journal of Theoretical Biology*, 18 (1968):280–315; "Developmental Systems Without Cellular Interactions, Their Languages and Grammars," *Journal of Theoretical Biology*, 30 (1971):455–84; and "Developmental Algorithms for Multicellular Organisms," *Journal of Theoretical Biology*, 54 (1975):3–22; and references.

40. Obviously, to get the picture of the leaf as given in the figure, one has to have a method of giving a visual interpretation to the mathematical symbolism of Lindenmayer's system. An example, simpler than that given in the text, will show precisely how such a model of development works (from Lindenmayer, "Algorithms"). The problem is to account for the generation of the gross form of a leaf. Since leaf cells originate at the leaf margin, it is the marginal row of cells that is important. Let us think in terms of five different cellular states, *a, b, c, d,* and *k,* and let the production rules from one state to the other (the "state transition function") be $a \rightarrow cbc$, $b \rightarrow dad$, $c \rightarrow k$, $d \rightarrow a$, $k \rightarrow k$. (For simplicity, two of the production rules stipulate that one cell gives rise to three cells. One can easily construct a similar but more complex system with only binary cell divisions. We assume also that the environment is constant, although functions can be devised to respond to environmental change.) Let *a* be our starting cell. We then get the developmental sequence in figure A.

a
cbc
kdadk
kacbcak
kcbckdadkcbck
kkdadkkacbcakkdadkk
kkacbcakkcbckdadkcbckkacbcakk
kkcbckdadkcbckkkdadkkacbcakkdadkkkcbckdadkcbckk
.
.
.

Figure A

In fact, from the fourth array on, we have a repeating structure.

$$S_n = k\, S_{n-3}\, S_{n-2}\, S_{n-3}\, k \qquad (n \geq 4)$$

Now, interpret *a* and *b* by sharp projecting tips, *c* and *d* by lateral lobe margins, and *k* by notches, and we obtain the leaf structures shown in figure B.

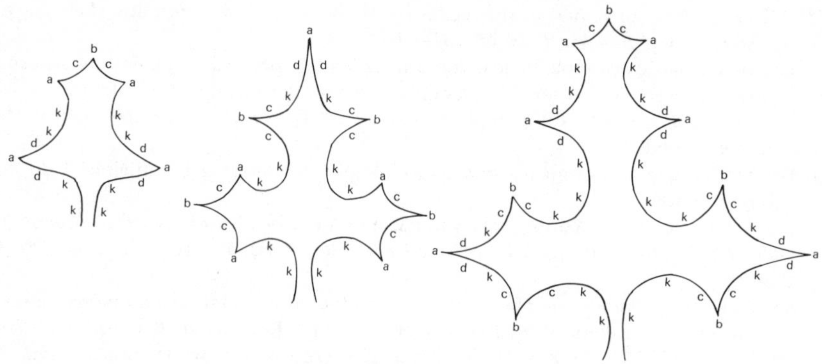

Figure B

Obviously these functions and constructions are not themselves axioms and theorems of a hypothetico-deductive system, but by making claims about them one can easily erect such a system. Moreover, one can find and prove "metatheorems"—claims about these claims about functions. One can, for example, show what kinds of functions can and cannot lead to what kinds of development.

41. M. Scriven, "Explanation and Prediction in Evolutionary Theory," *Science*, 130 (1959):477–82; T. A. Goudge, *The Ascent of Life* (Toronto: University of Toronto Press, 1961); J. J. C. Smart, *Philosophy and Scientific Realism;* G. G. Simpson, *This View of Life* (New York: Harcourt, Brace and World, 1963); B. Glass, "Explanation in Biology," *Journal of the History of Biology*, 2 (1969):47–54; E. Mayr, "Scientific Explanation and Conceptual Framework," *Journal of the History of Biology*, 2 (1969):123–28.

42. M. Ruse, "Narrative Explanation and the Theory of Evolution," *Canadian Journal of Philosophy*, 1 (1971):59–74; "Is the Theory of Evolution Different?"; *The Philosophy of Biology;* "Narrative Explanation Revisited," *Canadian Journal of Philosophy*, 4 (1975):529–33.

43. Much more discussion on this claim can be found in Ruse, "Is the Theory of Evolution Different?", and *The Philosophy of Biology*.

44. M. B. Williams, "Deducing the Consequences of Evolution: A Mathematical Model," *Journal of Theoretical Biology*, 29 (1970):343–85.

45. See M. Ruse, "Woodger on Genetics: A Critical Evaluation," *Acta Biotheoretica*, 24 (1975):1–13.

46. See P. J. Vorzimmer, *Charles Darwin: The Years of Controversy* (Philadelphia: Temple University Press, 1970); but see also M. Ruse, "The Darwin Industry: A Critical Evaluation," *History of Science*, 12 (1974):43–58, and idem, "Reduction, Replacement, and Molecular Biology," *Dialectica*, 25 (1971):39–72.

47. I am not sure how far one would get away from population genetics by doing something like this. It seems to me that even if one interpreted "biological entity" at the level of the organism, one would still need a fairly detailed knowledge of the principles of heredity in order to apply Williams's theory; for example, such knowledge would seem to be presupposed were one to claim that a certain group of organisms will expand in number at the expense of others by virtue of their peculiar characteristics. One must know something about the transmission of the characteristics.

48. See Ruse, *Philosophy of Biology*. See also idem, "Functional Statements in Biology," *Philosophy of Science*, 38 (1971):87–95; "Biological Adaptation," *Philosophy of Science*, 39 (1972):525–28; "Teleological Explanation and the Animal World," *Mind*, 82 (1973):433–36; "A Reply to Wright's Analysis of Functional Statements," *Philosophy of Science*, 40

(1973):277–80. I was converted to this claim by M. Beckner, "Function and Teleology," *Journal of the History of Biology,* 2 (1969):151–64.

49. Use of Fermat's principle of least time does lead to physical explanations where a cause is partially explained in terms of its effect, but unlike biological teleological explanations also the *explanans* (the thing explaining) contains the cause of the *explanandum* (the thing being explained).

50. For this reason, although I argue that biological teleology is irreducible, I do not think it to be irremovable.

51. For details, see R. M. Young, "Darwin's Metaphor: Does Nature Select?", *Monist,* 55 (1971):442–503; and C. Limoges, *La Sélection naturelle* (Paris: Presses Universitaires de France, 1970).

52. See D. L. Hull, *Darwin and His Critics* (Cambridge, Mass.: Harvard University Press, 1973); M. Ruse, "The Relationship Between Science and Religion in Britain: 1830–70," *Church History,* 44 (1975):505–22; "Whewell and the Argument from Design," *Monist,* 59 (1976); "Metaphysical Questions in the Darwinian Revolution," in *Pragmatism and Purpose* (Toronto: University of Toronto Press, 1977); and "Charles Lyell and the Philosophers of Science," *British Journal for the History of Science,* 9 (1976):121–31.

53. D. Brewster, *More Worlds than One: The Creed of the Philosopher and the Hope of the Christian* (London: Murray, 1854).

54. R. Owen, "On the Generation of the Marsupial Animals, with a Description of the Impregnated Uterus of the Kangaroo," *Philosophical Transactions* (1834):333–64.

55. W. Whewell, *Philosophy of the Inductive Sciences* (London: Parker, 1840), vol. 2, p. 78.

56. W. Whewell, *On Astronomy and General Physics: Bridgewater Treatise 3* (London: Pickering, 1833), p. 149.

57. Whewell, *Philosophy,* vol. 2, p. 86.

58. I discuss Darwin's use of the model of artificial selection in M. Ruse, "The Value of Analogical Models in Science," *Dialogue,* 12 (1973):246–53; "Charles Darwin and Artificial Selection," *Journal of the History of Ideas,* 36 (1975):339–50; "Charles Darwin's Theory of Evolution: An Analysis," *Journal of the History of Biology,* 8 (1975):219–41. I have more to say on Darwin's use of models in "The Nature of Scientific Models: Formal v. Material Analogy," *Philosophy of Social Science,* 3 (1973):63–80.

59. For details, see Ruse, "Darwin and Artificial Selection."

60. G. C. Williams, *Adaptation and Natural Selection: A Critique of Some Current Evolutionary Thought* (Princeton: Princeton University Press, 1966).

61. I think my analysis gives support to Max Black's analysis of metaphor (*Models and Metaphors* [Ithaca: Cornell University Press, 1962]). I argue that the use of the artifact metaphor in biology leads to a kind of understanding impossible without the metaphor.

62. An extensive discussion of this problem can be found in I. Scheffler, *The Anatomy of Inquiry* (New York: Knopf, 1963).

63. Modern biologists (like Mayr, *Animal Species*) argue vehemently that they have broken with biological typological thought. Let me therefore emphasize that I am *not* arguing for types in the sense of some Platonic Form in which all organisms of a taxon participate, which seems to be the thesis found objectionable. Rather, as will become immediately apparent, I allow for intrataxon (particularly intraspecific) differences.

64. If it is claimed that this statement is analytic because, by definition, genitalia help in reproduction, modify the statement by referring to the actual organs.

65. One might put forward this argument even though one agrees with what I argued earlier, namely that taken *literally* "x exists in order to do y" is explaining x in terms of a future y, not past y's. One might just feel that this consequence is so absurd that no biologist could want to be taken literally—his language is just a convenient shorthand for something else (a nonteleological something else).

66. That is to say, the functional analysis of x could not be explained in terms of

selection for previous *x*'s because of *y*-performance. One could, of course, give some causal explanation of *x*—perhaps *x* was selected for some different (that is, non-*y*) performance, or perhaps *x* was nonadaptive but pleiotropically linked to some other adaptive character.

67. G. G. Simpson, *The Major Features of Evolution* (New York: Columbia University Press, 1953), pp. 189, 194–95.

PETER MACHAMER
Ohio State University

Teleology and Selective Processes

> It is now possible to determine the sense in which the future is immanent in the present. The future is immanent in the present by reason of the fact that the present bears in its own essence the relationships which it will have to the future. It thereby includes in its essence the necessities to which the future must conform. The future is there in the present, as a general fact belonging to the nature of things. It is also there with such general determinations as it lies in the nature of the particular present to impose on the particular future which must succeed it.
>
> —Alfred North Whitehead
> *Adventures of Ideas*

The history of teleological terms and final causes in the writings of philosophers and scientists is long and checkered. Hardly had Aristotle worked out some of the details for speaking about final causes in naturalistic terms than Theophrastus devastatingly criticized their use with regard to metaphysics and first causes. Later, Christianity raised teleological concerns and final causes to pride of place, and extensive discussion and refinement was given them, culminating in over six hundred pages in the *Disputationes Metaphysicae* of Suarez. But within thirty-five years, Descartes (though he carried the Suarez text about with him from hiding place to hiding place) effectively argued against their use on theological grounds. Even Leibniz, for whom final causes were paramount, never succeeded in establishing that part of his program. In contemporary times, teleological concerns and final causes have come into the literature again, in the writings of contemporary biologists and philosophers of biology, in the work of some computer theorists, and in the musings of philosophers concerned with functionalist theories of mind.

In the present essay I want to examine a claim that seems to be held widely today but is little supported. It is the claim that by describing the process of selection that operates upon an individual (or species or subclan) one has provided sufficient grounds for talking about that individual (species or subclan) in teleological terms. In addition, I hope to show more clearly than the existing literature does how teleological talk is connected with explanations provided by different, though related, theories. Some of this ground has been worked by David Hull and Mary Williams, but I think not in the detail or clarity required.

A strong claim relating teleology and selection was made by William Wimsatt:

> Given the operation of a differential selection process, it is possible to show that any system resulting from this process has all the relevant logical features of purposiveness and teleology.[1]

Unfortunately, Wimsatt never shows what he says can be shown. David Hull takes himself to be repeating Wimsatt's claim when he writes: "He [Wimsatt] argues that a necessary requirement for a system to be teleological is that it arise, either directly or indirectly, through a selection process."[2] But Wimsatt only claimed that the operation of selective processes was sufficient for teleological talk. Hull himself seems to go on to claim sufficiency, when he says that teleological systems arise directly or indirectly, through selection processes.[3] Thus, both Hull and Wimsatt seem to claim that the operation of selective procedures upon a system is sufficient for that system's being teleological. Both Hull and Wimsatt, following Ernst Mayr and Mary Williams, argue that the way in which selection procedures legitimate such teleological talk is by defining goal states or preferable states of the system, at either the individual or species level.[4] Though Hull goes on to talk about hierarchical theories determining goal states and Wimsatt about hierarchical complexity, I find the connection between these notions and teleology unclear. I want to step back to basics and look tediously and pedantically at an argument (which they never give) that would take us from a premise concerning the operation of selection on a system to teleological talk about that system. The scrutiny of such an argument should allow for increased understanding of teleological systems and talk about them.

The claim to be examined is whether and, if so, in what way one can argue from a premise concerning a system's having been selected to a conclusion that attributes purposes, preferential states, needs, or desires to that system. In order that we know what counts as a teleological system, I shall assume that if any system is legitimately or correctly describable in terms of its needs, purposes, preferential or goal states, or desires, then this, plus the fact that it was selected, would be sufficient grounds for calling a system teleological. I think this way of stating the requirement will accord with everyone's intuitions concerning what is, at least paradigmatically, teleological. It is worth noting in this connection that traditionally to describe a system or set of causal processes as having a preferred output or goal state was taken as *the* criterion of a teleological system or set of processes.[5] That the output was not achieved or the goal not attained (Hull's criticism of this criterion) was not taken as negating the teleological character of the system or processes. The failure of attainment of its natural end was to be explained away by hindrances or interfering processes.

I spoke of systems correctly describable as having needs, or purposes, or desires, or goal states, or preferential states indiscriminately. Which of these descriptions is best for a given system or set of causal processes will depend upon the particular nature of what is being described. I feel quite happy about calling survival or nutrition a need of a system. I am much less happy about calling making love to blondes or becoming a tree the need of any system or process. The former would seem to be a desire of some men or women, while in the latter case it is true that certain kinds of seeds in some sense have the natural end of becoming certain kinds of trees. Thus, in what follows I shall stick with this disjunctive characterization of what correctly describes the system or process. All of these descriptive terms have one thing in common in that each is related in a peculiar way to kinds of ensuing responses or causal consequences from earlier states. Needs are satisfied, desires and purposes fulfilled, goals and preferences realized or attained. Any description of a system or process given in terms of such purposive characterizations carries with it a concomitant commitment to count or describe outputs or subsequent states in correlative ways.

I shall simplify what follows by speaking only about systems, and not also about causal processes. This is harmless, for any causal process can be modeled as a system of successive states, each of which is related to its successor and predecessor in specifiable ways. I will not argue for this here, but if there is disagreement, then let me have it for the sake of argument; or whenever I speak about systems, add in your mind "causal processes."

In speaking of systems one must talk about the systems themselves, their inputs, and their outputs. Descriptions of the systems will be in terms of mechanisms or internal structures that relate given kinds of inputs to specific kinds of outputs. Those internal states or structures are describable in terms of mechanical or functional parts. In the most completely described systems, the functional parts have associated with them independently identifiable entities or properties, that is, entities or properties which can be picked out independently of knowing the function they perform. In addition, one must be able to describe inputs and outputs. The input descriptions will often include reference to the larger environment in which the system is located and of which the particular input (often, in psychology, called the effective stimulus) is a part. The output or response will often feed back upon the input and upon the system itself and may cause the system to modify itself.[6]

This quick characterization is prerequisite to getting on with analyzing the argument concerning teleological systems in two ways. First, it is necessary to realize that whatever a selection procedure is, it is something that operates on individual systems (though these may be at different levels). Selection procedures must somehow bring about changes in existing individual systems and cause them to be different from the way they or their predecessors were. To say that a system was selected for is to say that it was selected for a certain property or trait. This property or trait must be a part of an individual system. Such parts I have called mechanisms or structures. Traditional examples of selected traits include color for animals in the case of natural selection, or structures allowing for new perceptual discriminatory abilities in either natural selection or learning cases.

The literature traditionally speaks of three different kinds of selective processes. The paradigmatic case is that of biological natural selection. The other two types include trial-and-error learning procedures and, derivatively, the setting of goals or giving of programs by third parties (or another system). This latter is said to be selection because the third party makes or creates the system in question according to a certain plan or goal. It is created to operate in a specific fashion.

The first two kinds of selective processes are those most commonly discussed. They operate on individual systems, but through different time scales. But even to say this is misleading. Natural selection may be properly said to operate upon species (or, following Mary Williams, on Darwinian subclans)[7] in the sense that fitness is said to be a property of subclans, not of the individuals that compose them. It is still true that the properties or traits selected are properties or traits of individuals that compose the subclans. Learning theory operates upon an individual over its lifetime. In both cases of selection, the environmental conditions surrounding a system during its lifetime causally affect it differently than they affect other similar systems. In the case of natural selection a given system is caused, for example, to reproduce at a greater rate than some other similar system because the environmental conditions are such that they favor it. In learning cases, for example, the prevalence in the environment of certain kinds of stimuli cause the system to learn to discriminate them effectively, while the absence of similar stimuli in other environments leaves similar systems without that ability. In many interesting systems, both kinds of selective processes are operative, and often (as the nature/nurture controversy has shown through the years) it is hard to distinguish which properties or traits of the system are due to which kinds of selective processes.

Following leads suggested by the philosophers of biology mentioned earlier, it is fruitful to expand these remarks by considering different levels (or kinds) of laws that are applicable in explaining the functioning of a given system. All of the philosophers speak of hierarchies of laws and have in mind two sets of laws or theories which correspond to the levels I have discussed: the evolutionary level, which gives laws of species or subclan selection, and a system level, which provides laws explaining how a given system works at a given time. This two-tiered model seems to me to blur important differences, especially concerning the second, or systemic, tier. I would suggest that there are at least three (possibly four) relevant levels of laws or explanations. First is the evolutionary level, which can be said to determine the system in certain quite general respects and to set limits to subsequent developments during its lifetime. Then, there are laws that explain paradigmatically how given kinds of systems function or develop. These laws describe the typical developmental sequences for kinds of systems. Finally, individual characteristics or traits of individual systems must be explained. This comes about in two ways: If one considered an individual and the particular idiosyncrasies it develops over a lifetime, then these would be explained in terms of deviations or additions to the paradigmatic schema described at the second level. So while all people typically learn visually to discriminate red or three-dimensional objects, only a few learn to discriminate the differences in color between two wines from different chateaux in Médoc. In addition to these lifetime individual differences, one must also be prepared to explain individual differences between systems at a single time. Immediately prior conditions can have an effect upon the discriminatory abilities of two different systems. Studies of set phenomena by Jerome Bruner and others have demonstrated this. These environmental causes do not modify the system permanently, only temporarily.

To focus my discussion, let me quickly go through some examples of these different levels. Consider the human visual system and its ability to see. For present purposes the system can be described as if it were independent of the other bodily systems. The visual system of a human is explicable from an evolutionary standpoint, showing how it is that we have come to be possessors of a chambered eye with a fovea and the consequent ability or need to scan, why it is that certain frequencies and not others are registered by the visual system, and in general in what ways the human system has proved to be adaptive to its long-term environment.[8] At the next level one could talk about more specific and transient environmental conditions and social pressures that surround individuals at given times or stages in their lives. At this level one explains why it is that the Zulus, raised in their circular culture, are not subject to the Mueller-Lyer illusion[9] and, more generally, how it is that people go through various developmental stages in their perceptual abilities. Explanations at this level are relying on societal or environmental pressures which act upon individuals during their lifetimes and which explain the various abilities of the group or kinds, not specifically determined by their evolutionary or genetic heritage. The paradigmatic form of explanation here is the kind to be found in good texts on the psychology of perception or its physiology. The third level of explanatory factors considers those environmental or psychological events which determine peculiarly individual lifetime traits and characteristics. At this level one explains deviations from, and fills in details of, the norms established earlier. One explains a particular individual's ability to recognize Cadillacs or to see what most others cannot. These explanations provide, to use a barbarous word, the personcentric characteristics of an individual's abilities. Finally, the fourth level provides explanations for peculiar abilities of individuals at a given time. These explain why Bob sees a zero instead of the letter *O*, when he has been shown a series of numbers.

Needless to say, these levels of explanation are not independent, and each makes use of and in varied ways depends upon the others. But in each of these cases the laws seem to explain various aspects of the system in much the same way. That is, the law tells us that certain kinds of events present in the organism's environment or the environment of his ancestors caused him to come to have specific traits or characteristics. These in turn are related to kinds of behavioral abilities the system has. These causing events (be they evolutionarily environmental, present environmental, social, or psychological) can be called selective pressures and act in such ways that they select the individual for various characteristics and abilities.

Having said all that I shall at present about systems, kinds of selective processes, and the levels of laws that govern both, let me turn to the argument I want to examine. Recall that I was interested in seeing whether the fact that selection was operative upon a given system was sufficient for the claim that that system was legitimately describable in terms of needs, purposes, desires, goals, or preference states. Having indicated that selective processes must operate in all cases upon individuals and in all cases select for some particular characteristic or trait of an individual, the first premise of a Wimsatt-Hull type of argument might be put:

(1) *S*'s having trait *i* were selected for under conditions *C* over *S*'s having *j* or not-*i*.

The reason for stating this premise in such a cumbersome way is that characteristics or traits are possessed by individual systems of kind *S*, and it is the environmental or social conditions that causally favor *S*'s having *i* over some alternative but similar kind of system, that is, *S*'s having *j* or not-*i*. Also, I have attempted to state the premise in a way that is general enough to allow *S* to be either a subclan or an individual of a kind; similarly, it should be general enough to allow the kind of selective procedure involved to be operative at any level. Thus, the selection of *S*'s having *i* will be explicable by the laws operating at a given level, where the levels correspond to those I spoke of earlier.

From this point it seems the argument might proceed in various ways. (One is unsure which way the proposers meant it to go, since they never gave it.) One could concentrate upon the trait *i* and attempt to describe it in more detail, exhibiting its structural characteristics. At this stage such a move would be premature, since we as yet have no indication what *i* was selected for. Alternatively, the argument might proceed by specifying the conditions *C* in more detail and showing how and why those conditions selected *i*. Thus, one might talk about the ecological environment of *S* in terms of the coloring of its flora, showing how *i* (a color property of *S*) allowed *S*'s having *i* not to be eaten by predators and, thus, showing how there were more *S*'s having *i* available to reproduce than *S*'s colored in other ways. But taking this line would require another aspect of the system, the output of *S*. It is better to refer explicitly to this component in a second premise. If we call the relevant output of the system ϕ, then the next premise might be

(2) *S*'s having *i* ϕ-ed.

But this premise will not be sufficient for continuing on to claim that *i* somehow fulfills (or tends to fulfill) a purpose, need, or desire (or achieves a goal state) for *S*. In speaking about selective processes I argued that a selective process was one which caused *S* to have an ability different from systems similar to *S*. Therefore, the second premise, which is still an assumption, must be somehow comparative between *S*'s having *i* and those not having *i*, so

(2′) *S*'s having *i* ϕ-ed better than having *S*'s having *j* or not-*i*.

I hope that (2′) can be established by a straightforward argument. If, for example, ϕ is taken as "reproduced," it could be established that *S*'s having *i*, a particular kind of coloration, were better at reproducing than those having a different coloration. Rather, it can be established if we know what "better at reproducing" means and we have a set of criteria for it. In this example it might be plausibly assumed that "better at reproducing" meant "reproduced at a relatively greater rate over a period of time t_1 to t_n." Notice that one has to import the criterion as to what counts as better ϕ-ing from somewhere else, that is, from an accepted body of beliefs, or sometimes from a theory. Here is the first point where it is clear that additional areas of belief or knowledge become involved in our argument and, thus, in our theorizing about teleological systems. From (1) and (2′) we can get

(3) *S*'s having *i* ϕ-ed better (or were better at ϕ-ing) under conditions *C* (which selected for *i*) than *S*'s having *j* or not-*i*.

So far it has been established by the argument that over some past period of time when conditions *C* obtained, *S*'s which had *i* came to be able to ϕ better than some similar *S* (not having *i*), where there are some accepted criteria for what counts as better ϕ-ing. At this point something else must be imported. We need to know something about ϕ-ing and what it does for *S*. If the argument is to yield needs, purposes, desires, and so forth as a conclusion, we need to know not just details as to how ϕ-ing is carried on by *S;* whether it lays eggs or has intercourse will not help us define a purpose, need, or desire for *S*. What is needed is information about ϕ-ing which fits ϕ-ing into a more general theory about *S*-like systems. An example will help here. If ϕ-ing is reproduction, then what will help us find purpose for *S* is some account of why reproduction is important for species or subclans. Generally, it might be said that greater reproduction is an important ability because it tends to promote the survival of the species or subclan. In general, we demand for our argument that

(4) ϕ-ing is important for *S*'s because it tends to promote ψ.

I used "tends to promote" in (4), but I could as easily have used words like "fulfill," "accomplish," or any one of a variety of success words. This is the point where Aristotelians traditionally spoke of "actualization." Moving quickly for a moment, it would seem that ψ in (4) needs further explication. Again, for example, survival is a need that species or subclans have if they are to continue to exist. Thus, it would seem that we need yet another premise telling us why ψ is a need, desire, purpose, or goal state for *S*'s. We need a premise like

(5) ψ is a need, purpose, desire, or goal state of *S*'s.

But this seems to assume as a separate premise what we were trying to prove from (1). If we are allowed (4) and (5) it would seem that we could easily get a conclusion of the kind desired, something like

(6) *i* serves a purpose ψ (fulfills a need, satisfies a desire, etc.) with respect to *S*'s better than *j* or not-*i*.

The question obviously is where one gets premises like (4) and (5). The answer, given the examples I have used, is from evolutionary theory.

Evolutionary theory tells us how to count what is better at reproducing than something else (sheer numbers, it says; the pleasure is irrelevant at this level for its purposes). Evolutionary theory tells us the significance of reproduction for survival of a species, and it tells us (though, perhaps, as part of the commonsense background of the theory) that species which do not continue to survive will cease to exist. We can justify premises (4) and (5) (and the criterion used in (3)) by bringing in another theory. Indeed, that another theory was indicated should have been obvious to us by looking at the "because" in premise (4), which literally tells us that ϕ-ing is causally related to ψ. Of course, we would need a theory to show us how.[10]

It might seem unclear whether the theory needed to justify our use of (4) and (5) is a different theory than the one which lies behind premises (1) through (3). I think it is. Premises (2′) and (3), about *S*'s with *i* being better at ϕ-ing, are established by considering various *S*'s during their lifetimes and seeing whether or not they ϕ better under conditions *C* than their competitors. No premises or laws from evolutionary theory are required to establish their truth. Either the same is true of premise (1), and only (4) and (5) require an additional theory, or, if not, then premise (1) is just another invocation of evolutionary theory, which needs to be supplemented by making explicit certain other aspects of the theory—to wit, those mentioned in (4) and (5). In any case there are at least two levels of theory operating in this argument. Minimally, it follows that the level of theory just dealing with natural selection is not sufficient for attributing purposes, needs, desires, or goal states to the system *S* and its kind. This follows because we need either premises (2) and (3) or (4) and (5) to get the desired conclusion, and these sets are from different theories. But there is a more interesting point to be made, even though my original question is answered.

Premises telling us why ϕ-ing or ψ is important or significant for *S* must be drawn from theories at a different level (or, perhaps, just a different body of theory) from those describing the operation of *S*, or the functioning of *i* for *S*. It is this other theory that exhibits the importance of ϕ by relating it to ψ and that allows us to make plausible the claim that ϕ tends to fulfill a purpose for *S*. If I can recast some of the claims I have made, what I have discussed is how selective conditions, *C*, operated upon *S* and where *S* had a mechanism or trait, *i*. Trait *i* allowed *S* to respond to *C* better than some alternative form of *S* that had non-*i* mechanisms or traits. The claim is that *i* is the mechanism (or part of the mechanism) by which *S* responds to *C*-type conditions and which causes *S* to issue an output different from alternatives. In most general form: Conditions *C* caused *S* to have different outputs than alternative *S*-type forms, and it did so by means of *i*, the mechanism by which *S* operates.

In this form it can be seen that all one can claim is that selection processes cause systems to issue in different responses (for example, behaviors or properties of offspring) when contrasted with other similar systems. Most of the time one is interested in those responses which are different in ways beneficial to the system or which fulfill needs, purposes, desires, and so forth of the system. Thus, the usual talk about fitness or adaptiveness. But there is no reason why one could not explain disadvantageous selection or even neutral output within the same scheme. The important point is that in every case, an additional theoretical claim is needed which explains why the differential output is significant (for good or ill) for the system in question.

As noted, the end or goal state for *S* which ϕ-ing tends to satisfy is specified by a theory or set of beliefs from a different level (or body of theory). In the above argument, premise (4) specified the relevant state, ψ, in the auxiliary theory, while premise (5) made explicit the claim that ψ was significant for *S*-type systems. In cases where ϕ is a criterion which is significant for *S* because it allows it, for example, to survive or avoid harm or gain pleasure, the role of the auxiliary theory is often obscured. It is obscured because that *S* needs to survive, or avoid harm, or gain pleasure, seems painfully obvious. Thus, to require that we have a theory which explicitly lays out these things might seem to be pedantry of the worst sort. But if we notice that we do not have the conceptual capacity to specify harm-avoidance, pleasure-attainment, or survival-benefit within our theory of *S* (as its inputs, internal states, and outputs), we can see that something must be added even to state the claim about the role ϕ plays for *S*'s via ψ.

One can be asked to justify these appeals to an additional body of theory. Specifically, one can be asked to justify his use of such premises as (4) and (5). Ultimately the justification of such premises as (5), which claim that a certain state is a need, goal, and so forth, will be dependent upon the justification of the theory or set of beliefs in which those claims occur. They will be justified by showing how the theory which makes the claims is preferable to its competitors or by appealing to the accepted background beliefs of the community. The interesting concern of moment is more specific: The claim that ψ is a goal state, purpose, need, desire of *S* must be justified. This concerns how it is that such a claim fits into the theory of which it is a part and not just a claim about the general adequacy of that theory.

To exhibit how it is that ϕ, which is an output of *S*, promotes ψ, and why it is that ψ is significant, it is enough to show what is the role of ϕ in the life of *S*'s (over a lifetime or over evolutionary time), how this role of ϕ relates to those states described by ψ, and how the states ψ are related to other states of *S*-type systems, to activities (outputs) of those systems, and to inputs or conditions of the systems. Thus, ψ becomes a descriptive or explanatory variable of a more general system, for example, of *S*'s in evolutionary time. The use of such variables is justified, or seen to be useful, by exhibiting their connections with the other explanatory elements at that level.

In the evolutionary example used above, better reproduction ability was important ultimately because it tends to promote the survival of the species (an instance of premise 4). The significance of the survival of a species is to be explained, for example, in terms of the influence which that species has upon its environment (or upon the ecological niche it inhabits), its influence over time upon other species that are spatially proximate to it, its role in the food chain, and so forth. All of these other factors show us why survival of *S*-type systems is significant, and allow us legitimately to speculate about changes that might occur if these were not extant. They spell out for us why *S*'s survival is important or an important precondition for the role *S*'s play, and they lead us back to asking how and why it was that *S* survived, which is what is answered in part by the antecedent of premises like (4) and premises (2) and (3) (concerning *S*'s better reproduction abilities). If this account is along the right lines, then talking about need, purpose, desire, or goal in premises such as (5) would seem to be shorthand ways of talking about either the role that ψ plays for *S*-kind of systems, or a necessary precondition (in the case of a need) for *S*'s having such roles. The roles are what is specified by the set of laws or beliefs which link ψ to other explanatory variables at this different level of theory. Finally, since either the best theory or the accepted communal beliefs can change, and since the conditions *C* which selected for *i* and under which ψ is significant for *S*-type systems can change themselves, there is no ultimate rationale for showing that *i* will always be good for *S* (or better than alternatives) or showing that ψ is significant for *S*-type systems. Given changes, what is significant and what is taken to be significant may change.

This procedure for specifying goals and the like may seem circular, but I think it is not. If we ask why a given trait of the system is significant for it, we can answer in terms of what it allows the system to do better than alternatives. If asked why that ability or output is better, we can justify it by calling attention to the conditions in which the system is at the time (or over a long time) and by showing how *given* those conditions, *if* the system is to survive or lead the good life or accomplish whatever end, it *ought* to develop such and such kinds of abilities. If asked ultimately why the system ought to survive or why the good life includes wines and truffles, we are going to have to justify those claims by contrasting them with their alternatives and arguing in terms of now accepted standards or theories that this alternative or theory is superior to others.

To illustrate these claims with a biological example, let me borrow from Mary Williams's comments about David Hull's book:[11] In explaining the phenomenon of sexual reproduction, one might ask, What is the effect of sexual as opposed to asexual reproduction? The answer will be a comparison between the two kinds of reproduction plus the demonstration or provision of a reason why one is superior; for example, sexual reproduction produces greater genetic variability within each generation. If asked why greater genetic variability is significant for organisms, the answer comes from another theory (in this case evolutionary theory), which tells that greater variability increases the probability of survival in a changing environment. In this case, again following Mary Williams, the evolutionary theory explains why greater variability is a preferable output of the system. It does so by linking it to an accepted claim, that systems need to survive. If we look now to a perception case, and ask why the frog has moving-edge detectors, we can explain that ability in terms of the frog's having dragonflies as its food, and then explain the latter in terms of evolutionary theory which specifies the frog's ecological position and its need for finding food in those kinds of circumstances. In social learning cases, we can explain a person's ability to form tannin taste sensations of the good wines in terms of the arduous training the person went through, contrasting young wines with old and noting differences of taste. If asked why this discriminative ability, we would respond in terms of a higher need, desire, or purpose, perhaps that of living a life with maximum pleasures, and this in turn would be justified by showing how it is that lives without good wines (and their pleasures) are inferior to those with them. In all these cases, the justification of why the trait or consequent ability is significant depends upon specifying the conditions under which it operates, and upon showing how the trait, the ability, and the conditions are linked together according to some theory or set of beliefs that we have. In all cases it is possible to call into question the adequacy of the linkages established. Reformers in science and society do just that.

I want to close this chapter with a final speculation concerning the conditions under which it seems most natural to use teleological concepts. We seem to use teleological concepts when we attempt to explain how and why what were heretofore considered different systems (or what were heretofore counted as different systems) are connected or related, and to explain how the one system has its effects (and conversely) upon the others. If it is possible for us to answer all the questions we have (at a time) without going beyond the intrasystemic knowledge we have about a given system and its works, there is no place for teleological concepts in our explanation. Thus, in such cases we feel odd about talking about selection, purposes, and their like. When we want to see the interrelation between two systems, then we need to make use of teleological concepts to explain those interactions. This just comes to saying that if for present purposes we can consider a given system as closed, then we do not need, for those purposes, to develop higher laws explaining the functional interdependence between systems. When systems are relatively unchanging, it is easy to view them as closed. If the kinds of conditions to which a system responds are undergoing change, having an adequate theory requires us to take into account the patterns of changing kinds of conditions. The changing conditions can come to be counted as another system, and they need to be described in such a way that we can see the connection between them and their changes and the way our original system operates and changes.

A caveat is needed at this point. I am not talking about intertheoretical relations in a way in which they have been traditionally spoken about, for example, as reduction. I do believe that many of the problems of theory reduction which are commonly raised are better viewed as questions about how systems are interrelated, rather than as how one system is reducible to another. My point at present is not to talk about such relations in any detail, but merely to note the interesting fact that when Leibniz, Hegel, and the other teleologists of the past invoked final causes or other teleological elements, they did so within the schema I have just described. All were trying to relate different levels of theory, for example, God and creature, historical evolution and individuals at a time, and so forth. Perhaps this fact, plus the insight that no theory can really be closed (or stand alone) will help to open the way once again for the use of teleological terminology and final causes in philosophy and scientific explanation.

NOTES

I should like to thank James Bogen, David Hull, Ted McGuire, Marshall Swain, and Mary Williams for their comments and their part in various discussions concerning aspects of the topics treated in this essay. Also I should like to thank those who participated in the discussion at Pittsburgh when a draft of this essay was read.

1. William C. Wimsatt, "Teleology and the Logical Structure of Function Statements," *Studies in History and Philosophy of Science,* 3 (1972):16.
2. David Hull, *Philosophy of Biological Science* (Englewood Cliffs, N.J.: Prentice-Hall, 1974), p. 113.
3. Ibid., p. 115. In correspondence, David Hull assures me that the sufficiency claim was not meant as logical sufficiency; only a contingent relation was suggested.
4. Mary B. Williams, "Deducing the Consequences of Evolution: A Mathematical Model," *Journal of Theoretical Biology,* 29 (1970):343–85; "The Logical Status of the Theory of Natural Selection and Other Evolutionary Controversies," in *The Methodological Unity of Science,* ed. M. Bunge (Dordrecht, Holland: Reidel, 1973), pp. 84–102; and in conversation. Mary Williams attributes the original insight, and not all the details, to Ernst Mayr.
5. I think this is basically the traditional criterion for teleology which was used by Aristotle and his followers in their theories concerning final causes and natural substances.

During the seventeenth century this theory became transformed into talk concerning preferred states associated with *conatus* (endeavors) and the laws of nature or motion that specified them. It is this recognizable aspect of the traditional role of final causes in laws and explanations which becomes lost in the post-seventeenth-century period.

6. Ultimately what is required are sets of descriptions for the input, system, and output which are correlative to one another in the ways in which causal descriptions are. Though this notion of correlative descriptions is rather vague at present, it can be exemplified by the relation that holds between the description of an effect as a puncture wound and the limitations which that description places upon possible causes of that effect, that is, the cause must be some sort of thin, sharp instrument (because only such can cause punctures). Cf. N. R. Hanson, *Patterns of Discovery* (Cambridge: Cambridge University Press, 1958), chap. 3; and William Ruddick, "Causal Connection," in *Boston Studies in the Philosophy of Science,* 4, ed. Robert S. Cohen and Marx W. Wartofsky (Dordrecht, Holland: Reidel, 1969), pp. 419–41.

7. Williams, "Deducing the Consequences of Evolution."

8. For this evolutionary approach to vision see J. J. Gibson, *The Senses Considered as Perceptual Systems* (Boston: Houghton Mifflin, 1966), chap. 9.

9. Richard Gregory, *Eye and Brain* (London: Weidenfeld and Nicolson, 1966), pp. 161 f.

10. It is worth pointing out that the approach taken here considers the problem as one of specifying the ways in which two different theories are related to each other. Seen in this way the problem is one aspect of the more general problem of specifying intratheoretical relations. Other aspects are more familiar, for example, reduction of theories or the use of auxiliary theories in explanations. The fruitfulness of conceiving things in this way lies in the fact that by so doing much of the literature on theories and their relations to each other becomes relevant to the problem discussed here. In addition, this perspective provides a genus for discussing a whole range of problems in the philosophy of science which were heretofore treated as separate. See the closing part of this chapter for further speculations on this aspect.

11. An unpublished letter concerning Hull's book, from Mary Williams to David Hull.

KENNETH F. SCHAFFNER
University of Pittsburgh

Reduction, Reductionism, Values, and Progress in the Biomedical Sciences

> Kant, with the Newtonian persuasion of the validity of mechanics, was sympathetic with the mechanical theories of biology, but he also discerned in the living organism the operation of reciprocal causes that empirically conformed more closely to the concepts of means and ends than to the concepts of mechanics; in fact, he went so far as to repeat emphatically that not even a blade of grass could be conceived to have been produced by the action of mechanical causes alone.
>
> —Scott Buchanan
> *Truth in the Sciences*

1. INTRODUCTION: REDUCTION, REDUCTIONISM, AND PROGRESS

The terms "reductionist" and "reductionism," as employed in scientific and philosophical literature, have a number of varied meanings associated with them. In the biomedical sciences—the focus of the present essay—a "reductionist" often means a scientist (or a philosopher) who believes that biological entities are "nothing but" aggregates of physicochemical entities. A related but different meaning of this term refers to a person who believes that only molecular methods and descriptions can yield sound scientific generalizations.[1] The first view may be characterized as ontological reductionism and the second, as methodological reductionism.[2] There are still further variant connotations associated with the term reductionism, such as that which tends to identify reductionism with a type of mental attitude somewhat similar to that found in *critical analytical* thinking. This last interpretation has been held, for example, in an extreme form, by T. Roszak, who has written:

> I use the word "reductionism" here broadly to designate that peculiar sensibility which degrades what it studies by depriving its subject of charm, autonomy,

dignity, mystery. As Kathleen Raine puts it, the style of mind which would have us "see in the pearl nothing but a disease of the oyster."

Abraham Maslow has characterized reductionist intellect as a "cognitive pathology" usually born of fear, rather like a kind of exorcism needed by the chronically distrustful to drive off the menacing spirits they sense about them in the world. Doubtless in our culture, reductionism traces back to the Judeo-Christian mania for desacralizing nature. It is mixed, too, with a compulsively masculine drive to demonstrate toughness, expel sentiment, and to get things under heavy-handed control. The result is a dismal conviction that knowledge can only be that which disenchants and cheapens both knower and known.[3]

A similar view also appeared recently in an article by E. J. Cassell, a clinical professor in the department of public health at Cornell, who distinguishes between analytic thought in "the reductionist mode" and "valuational thought," which is in an "integrative, synthetic, or constructionist mode." Cassell also believes that "analytic thought in medicine not only tends to drive out valuational thought . . . but also to exclude humanistic considerations from modern medicine."[4]

As a popular and current cultural anthropologist, Roszak may perhaps be thought somewhat removed from the biological sciences, and Cassell might be construed as requesting that we focus more attention on the purely humanistic, nonscientific aspects of medicine. The views of these authors, however, are not too dissimilar from those held by T. Dobzhansky, the late eminent geneticist and evolutionist, who argued squarely in the context of biological, scientific thought that:

there are two approaches to the study of the structures, functions, and interrelations of living beings—the Cartesian or reductionist and the Darwinian or compositionist. This does not mean that some biological sciences are reductionist and others compositionist, or that there are Cartesian and Darwinian phenomena. Biological phenomena do have, however, Cartesian and Darwinian aspects. Some biologists view their subject more from the reductionist and others from the compositionist side, and some are more adept at using Cartesian and others Darwinian methodologies.

Descartes considered living bodies, including human bodies, to be automata, i.e., machines describable in physical, and eventually, in mathematical terms. They may, accordingly, be studied by means of the famous Cartesian method. Like any complex phenomena, the phenomena of life must be divided into the simplest components amenable to study. . . .

Whether reductionist explanations are by themselves sufficient may, however, be questioned. Warren Weaver . . . wrote that "A person usually considers a statement as having been explained if, after the explanation, he feels intellectual comfort from reductionism alone. . . ."

To feel "intellectually comfortable" about our understanding of biological phenomena, Darwinian, or compositionist, explanations are required. The matter can best be stated in Simpson's words . . . "In biology then, a second kind of explanation must be added to the first or reductionist explanation made in terms of physical, chemical, and mechanical principles. This second form of explanation, which can be called compositionist in contrast with reductionist, is in terms of adaptive usefulness of structures and processes to the whole organism and to the species of which it is a part, and still further, in terms of ecological function in the communities in which the species occurs."[5]

What we encounter in these writers, from Roszak to Dobzhansky, is a distrust of a unified analytical approach to living systems and their interactive (including social) activities. The authors' views should not be mutually identified, but the thread of looking for something more, for some means of comprehension which is nonreductionistic, is common to the three.

It is the purpose of this essay to attempt to disentangle some of these often conflated meanings of reductionism. Specifically, I shall first define what I mean by the core sense of the reduction of one theory (or branch of science) to a different theory (or branch of science). This involves utilizing some philosophical terminology and logical machinery to construct an ideal of reduction. This ideal is seldom met but serves as a touchstone for analyzing and discussing actual "reductions"—most of which are, for reasons to be developed shortly, partial or incomplete.

Second, I shall argue that the reasons why reductions are almost always incomplete are dependent on two factors: (1) the dynamics of science in which theories are constantly evolving and being replaced, and (2) the goals or values of science which are not what I term "directionally reductionistic," but are rather, to use a clumsy word, "intelligibilistic." In very general terms—to be made more specific later—science, including biology, aims at understanding, at precision, and at control over nature, and only *accidentally* at directional reduction. This state of affairs also holds in molecular biology, a notion I have previously defended under the rubric of a "thesis of peripherality of reductionism."[6] A clearer comprehension of the values of science is accordingly required for understanding the nature of incomplete reductions in science.

Finally, I shall contend that the values at which science most directly aims have to be seen within a *broader valuational context* if we are to understand real scientific change and the shifting fortunes of reductionistic programs. This broader value penumbra reflects the values which judge both the psychological and social worth of specific scientific research projects and, in the most general form, the value of science itself as a human enterprise. I shall examine this thesis specifically within the biomedical sciences, which necessarily and explicitly involve certain general and fundamental human values such as "self-determination," inasmuch as biomedical research cannot progress without the use of human subjects.[7]

I shall argue briefly that, within the framework of this analysis, a person who subscribes to the general reduction-replacement model can nonetheless urge the pursuit of nonmolecular interlevel theorizing and experimentation in the biological sciences and can champion analytical but value-conditioned and value-laden thinking in science.

This essay therefore moves from a *prima facie* value-free reduction model to an analysis of values internal to scientific inquiry in an attempt to clarify the differences between accomplished reductions and the dynamics of scientific progress. The essay further suggests that a full analysis of scientific progress requires comprehension of both internal *and* external factors (including values). The article concludes by arguing that a unified analytical approach can treat both descriptive and normative aspects of science and also extrascientific experience. This unified analytical approach, which is scientific methodology broadly applied, is proposed as a broadly "reductionistic" alternative to those species of antireductionistic thought recommended by Roszak and Cassell.

2. REDUCTION: THE GENERAL REDUCTION-REPLACEMENT MODEL

Reduction in the core sense in which it will be examined in these pages means the explanation *or* replacement of one scientific *theory* or *branch of science* by another. A branch of science will be presumed exhausted either by its aggregate of theories, or by its theories and/or its *domain,* a term to be introduced and utilized further. I will therefore be using "reduction of a science" and "reduction of a theory" interchangeably. If the theory to be reduced is deemed adequate, that is, if it does not conflict with the prospective reducing theory and if the prospective reducing theory organizes the older theory's subject area and accounts for experimental results as well as or better than the theory to be reduced, then *explanation* of the reduced theory is the usual situation. If the prospective reducing theory is in clear conflict with the older theory, then the reducing theory, being by supposition of greater scope and precision, *replaces* the previous theory. In this latter case we have two prima facie distinct aspects of scientific advance intermingled: reduction (in the traditional sense) and theory competition. In the present essay I will initially deal with the case of reduction without conflict and later complicate the discussion to consider reduction with conflict, as it is the more usual one.

Reduction without conflict has been described essentially correctly by E. Nagel in his pioneering essay of 1949,[8] later expanded into a broader treatment of the issue.[9] Nagel proposed two formal conditions:

1. Connectability: in which "assumptions of some kind must be introduced which postulate suitable relations between whatever is signified by 'A' [a term appearing in the reduced science but not in the reducing science] and the traits represented by theoretical terms already present in the primary [or reducing] science."
2. Derivability: in which "with the help of these additional assumptions, all the laws of the secondary [reduced] science, including those containing the term 'A' must be logically derivable from the

theoretical premises and their associated coordinating definitions in the primary discipline."[10]

This model of reduction represents a most important first step in clarifying our thoughts concerning reduction. It needs additional development on several fronts, however, in order to clarify the nature of the logical structure of reductions, and also to bring the model into better accord with actual cases of reduction in science.

As Nagel pointed out in his earlier writings, the nature of the connectability assumptions admits of a variety of interpretations. He suggested the possibilities of logical connections, conventions, and factual or material connections (that is, physical hypotheses) as interpretations.[11] Discussion of the logical and epistemic nature of the connections in the literature has tended toward construing them under the last rubric and interpreting them as *synthetic identities.*[12] (The synthetic sentences might at some later stage of science be taken as definitions, but this move would not vitiate the synthetic character of the connections in the original reduction.)

A more difficult problem was raised by criticisms of Nagel's model of reduction by P. K. Feyerabend.[13] Feyerabend pointed out that Nagel's condition of connectability was never realized in practice and that successive physical theories were either mutually inconsistent or possibly even "incommensurable." For example, he noted subtle shifts from phenomenological thermodynamics to statistical mechanics, which made the laws of phenomenological thermodynamics *false* when taken in a strict sense. In addition, the *concepts* of phenomenological thermodynamics could not be associated in any simple, true sense with the concepts from statistical mechanics. Entropy, an important though nonprimitive concept, always either increased or remained constant in the "reduced" theory but could decrease in the "reducing" theory.

One possibility that suggested itself in an attempt to outflank this objection was to alter the reduced theory so that a modified reduced theory could be connected with the reducing theory.[14] This corrected reduced theory, symbolized as a T_2^* variant of a to-be-reduced T_2, might be connected with a reducing theory T_1. The possibility that the *reducing theory* also might require change to effect the reduction, that is, a change from T_1 to a T_1^*, was also a realistic one. These conditions can be summed up in what I have termed the general reduction model:[15]

General Reduction Model

T_1—the reducing theory

T_2—the original reduced theory

T_2^*—the "corrected" reduced theory

Reduction occurs if and only if:

(1) All primitive terms of T_2^* are associated with one or more of the terms of T_1, such that:

(a) T_2^* (entities)=function [T_1 (entities)]

(b) T_2^* (predicates)=function [T_1 (predicates)]

(This is the condition of referential identity.)

(2) Given fulfillment of condition (1), that T_2^* be derivable from T_1 supplemented with (1)(a) and (1)(b) functions. (This is the condition of derivability.)

(3) T_2^* corrects T_2, that is, T_2^* makes more accurate predictions.

(4) T_2 is explained by T_1 in that T_2 and T_2^* are strongly analogous, and T_1 indicated why T_2 worked as well as it did historically.

I now believe that even this model must be further generalized, however, if it is to encompass all that scientists traditionally have called reduction. Specifically, as suggested above, it must allow for those cases in which the T_2 is *not* modifiable into a T_2^* but rather is *replaced* by T_1 or a T_1^*. Though it is not historically accurate, a reduction of phlogiston theory by a combination of Lavoisier's oxidation theory and Dalton's atomic theory would be such a replacement. Some of the general logical conditions under which a theoretical conflict leads to selection of one theory and rejection of the other will be considered in section 3. For this section it will be assumed that partial or complete rejection of a theory to be reduced has occurred and the reducing theory has been accepted.

In the simplest replacement situation we have the essentially *experimental arena* of the previous T_2 (but not the theoretical premises) *directly* connected via new correspondence rules associated with T_1 (or T_1^*) to the reducing theory. A correspondence rule is here understood in the sense that I have discussed in previously published papers, specifically that it is a telescoped causal sequence linking (relatively) theoretical processes to (relatively) observable ones.[16] Several of these rules would probably suffice, then, to allow for further explanation of the experimental results of T_2's subject area by T_1 (or T_1^*). In the more complex but realistic case, we also want to allow for the possibility of a partially adequate *component* of T_2 being maintained together with the entire *domain* (or even only *part* of the domain) of T_2, to use a technical term first introduced in this sense by D. Shapere.[17] Thus, there arises the possibility of a *continuum of reduction relations* in which T_1 (or T_1^*) can participate. (In those cases where T_1 *or* T_1^*, in the exclusive sense of "or," is the reducing theory, I shall use the expression $T_1^{(*)}$.) To allow for such a continuum, T_2 must be construed not only as a completely integral theory but also as a theory dissociable into individual assumptions, and also associated with an experimental subject area(s) or *domain(s)*. The sense we give to 'domain' is that of a complex of experimental results which either are accounted for by T_2 or (where "or" is to be taken in the inclusive sense) should be accounted for by T_2 when (and if) T_2 is or becomes completely and adequately developed. The general reduction model, then, is to be modified into the general reduction-replacement model, characterized by the following conditions:

General Reduction-Replacement Model

T_1—the reducing theory

T_1*—the "corrected" reducing theory

T_2—the original reduced theory

T_2*—the "corrected" reduced theory

Reduction in the most general sense occurs if and only if:

(1)(a) All primitive terms of T_2* are associated with one or more of the terms of $T_1^{(*)}$, such that:

(i) T_2* (entities)=function [$T_1^{(*)}$ (entities)]

(ii) T_2* (predicates)=function [$T_1^{(*)}$ (predicates)]

or

(1)(b) The domain of T_2 be connectable with $T_1^{(*)}$ via new correspondence rules. (Condition of generalized connectability.)

(2)(a) Given fulfillment of condition (1)(a), that T_2* be derivable from $T_1^{(*)}$ supplemented with (1)(a)(i) and (1)(a)(ii) functions.

or

(2)(b) Given fulfillment of condition (1)(b) the domain of T_2 be derivable from $T_1^{(*)}$ supplemented with the new correspondence rules. (Condition of derivability.)

(3) In case (1)(a) and (2)(a) are met, T_2* corrects T_2, that is, T_2* makes more accurate predictions. In case (1)(b) and (2)(b) are met, it may be the case that $T_1^{(*)}$ makes more accurate predictions in T_2's domain than did T_2.

(4)(a) T_2 is explained by $T_1^{(*)}$ in that T_2 and T_2* are strongly analogous, and $T_1^{(*)}$ indicated why T_2 worked as well as it did historically.

or

(4)(b) T_2's *domain* is explained by $T_1^{(*)}$ even when T_2 is replaced. The (italicized *or*s should be taken in the weak, inclusive sense of "or.")

Such a model has a limiting case what I have previously characterized as the general reduction model,[18] which in turn yielded Nagel's model as a limiting case. The model also permits partial reduction/

replacement cases, where a partial reduction is understood in the sense introduced earlier. In addition, I believe that the general reduction-replacement model neatly meets the criticism directed at the general reduction model by Michael Ruse and David Hull that it could not distinguish between reduction and replacement.[19] It does so by accepting the criticism and including it in a general model which allows for a *continuum* of cases in which partial reduction and partial replacement can nonetheless be characterized as *reduction* in the most general sense.

It would be possible to develop the nature of the connections between T_1 (or T_1^*) and T_2^* in logical detail and apply them to the biological sciences, but I have done this elsewhere.[20] It is also unnecessary to develop the notion of a new correspondence rule and its logical analysis in these pages, as I have done so reasonably specifically elsewhere.[21] It is, however, desirable to make a brief digression to indicate how reduction of a domain can occur without the reduction of the (complete) theory. To do this we consider very briefly the cases of the replacement of the demonic theory of disease by the Galenic theory of disease, the reduction via replacement of Newtonian classical optics by the Young-Fresnel wave theory, and the partial reduction of nineteenth-century ether-theoretic optics by a relativistically construed Maxwellian theory of electrodynamics.

It has been a characteristic of many ancient and primitive cultures to understand diseases as being caused by malevolent spiritual beings. A popular demonic theory of disease also surfaced in the Middle Ages. A nonspiritual theory of disease etiology would not wish to identify its entities with demons or aspects of demons, but might well accept an aggregate of the *observable symptoms,* or the *syndrome,* from the previous demonic science. Aspects of the syndrome thus might be directly connected with results of hypothesized biological processes, for example, a disease caused by an unbalance of the four humors in Galenic medicine.[22]

A more modern and specific example of the direct connection of a new theory with the older observable domain of the replaced theory is the case of the replacement of the emission or corpuscular theory of light, championed by Isaac Newton, by the wave theory of light developed by Thomas Young and Augustin Fresnel in the nineteenth century. Reflection phenomena, refraction experiments, (including low-level generalizations such as Snell's law), interference and diffraction patterns, and polarization experiments were explained or partially explained by Newton's theory.[23] Almost all of these explanations were rejected by Young and Fresnel, who proposed different theoretical processes that were associated directly—without going through the previously rejected theory—with observable results such as light and dark patterns.

The experimental domain of the previous theory is explained, and in the *broad* sense the previous theory's domain is reduced, though the theory *per se* is rejected.

These two cases are cases of extreme theory rejection. There are other cases, however, in which, though it is unclear that a $T_2{}^*$ can be formulated, large aspects of a previous T_2 are absorbed in a reducing theory $T_1{}^{(*)}$. A historical case with this intermediate property would be nineteenth-century ether-theoretic optics, which was partially reduced by Maxwell's electromagnetic theory and then had one of its central assumptions—the ether—eliminated by the special theory of relativity. In that case *portions* of optical *theory* and the theory's explanations of experiments are preserved and modified in the context of a relativistic construal of Maxwell's electromagnetic theory. Fresnel's explanation of interference patterns would be a case in point.[24]

To this point I have treated reduction in terms of explanation and/or replacement but without introducing a *directional* aspect of reduction which is traditionally very important in reductions involving biology. The directional aspect is sometimes articulated in terms of "levels" in which reduction is assumed to involve an explanation or replacement *of a higher-level theory by a lower-level theory.* An example would be the explanation/replacement of the traditional theory of Mendelian-Morganian genetics by a molecular biochemical theory involving DNA and proteinaceous entities. The general reduction-replacement model as outlined above does not *per se* involve this "level" directionality, and it is thus capable of encompassing what might be termed unilevel reduction relations, such as can be found in the relation of the optical ether-theoretic and electromagnetic ether-theoretic theories in the nineteenth century.

In the biomedical sciences, however, the term "reduction" is often taken to have a directional connotation inherently associated with it. I believe this is a mistake and that it confuses our understanding of scientific change and progress. I will examine this issue in more detail in the next section.

It would be useful to point out here, however, a closely associated problem for the reduction-replacement model presented above, namely, that it may be *historically* misleading, even though it has a dynamic element built into it in its starred theories. Reductions, no matter where they may lie on the continuum from reduction as explanation to reduction as replacement, rarely if ever go to *completion.* The model presented above, with the exception of its explicitly *partial* aspects, is an ideal of *finished* science, of science that *has* evolved but has stopped evolving. The science we know today and the science we know from history do not have this characteristic.

Furthermore, it is important to note, as will be developed in the next section, that the reason why reductions are not completed follows from an intrinsic and central component of scientific inquiry. Reductions are not completed because science is constantly changing, and because reductions in a directional sense are only *accidentally* the aim of science. Let us now turn to these issues.

3. REDUCTIONISM AND VALUES IN SCIENCE

An examination of a number of historical episodes in the development of important new models and theories in the biomedical sciences discloses that reductions, in the directional sense, are not a primary goal, aim, or value of even *molecular* biology. The notion of a goal (or goals) in science also raises the interesting problem of value-free science. It is to the first of these problems, directional reduction as a goal, that I turn in the present section. The succeeding section addresses the value-free issue.

Let us consider briefly two historical cases of theory development in the biomedical sciences. The first, the Watson-Crick model of DNA (figure 1) is often taken as a paradigm case of reduction in biology.[25] Here in a chemical model, with its component parts completely specified in chemical terms, is the molecular basis of the gene and the chemical code for all parts of a living organism.

I would argue, however, that the molecular model was not primarily important for its chemical aspects but rather because it was easy to see, to quote J. D. Watson and F. H. C. Crick, "that [the model] immediately suggests a mechanism for its self-duplication," and also that it was not difficult to envision how the sequence of nucleotide bases could be "the code which carries the genetical information."[26]

Both of these aspects of the model were most important *biologically,* but the mechanism of self-duplication is *still not completely* understood in physicochemical terms. Alternative models have been and continue to be proposed, but the source of the uncoiling energy, the exact mechanism of complementary pairing, and the enzyme(s) involved, are still unclear.[27] (For example, the Kornberg polymerase I, which seemed responsible for replication, is now thought to be a DNA repair enzyme.) The account of self-duplication is, then, physicochemically incomplete. Nevertheless, the Watson-Crick model has had a most important influence in suggesting experiments and clarifying the revisions of genetics, necessitated by the fine-structure analyses of the gene. The Watson-Crick model is also part of the background of an elaborate theory of protein synthesis which constitutes one of the most powerful theories of molecular biology. One point that I would like to make on the basis of this example is that physicochemically incomplete models and theories which significantly aid in the reduction of classical biology, in this case classical genetics, can still play a most important role in conditioning the

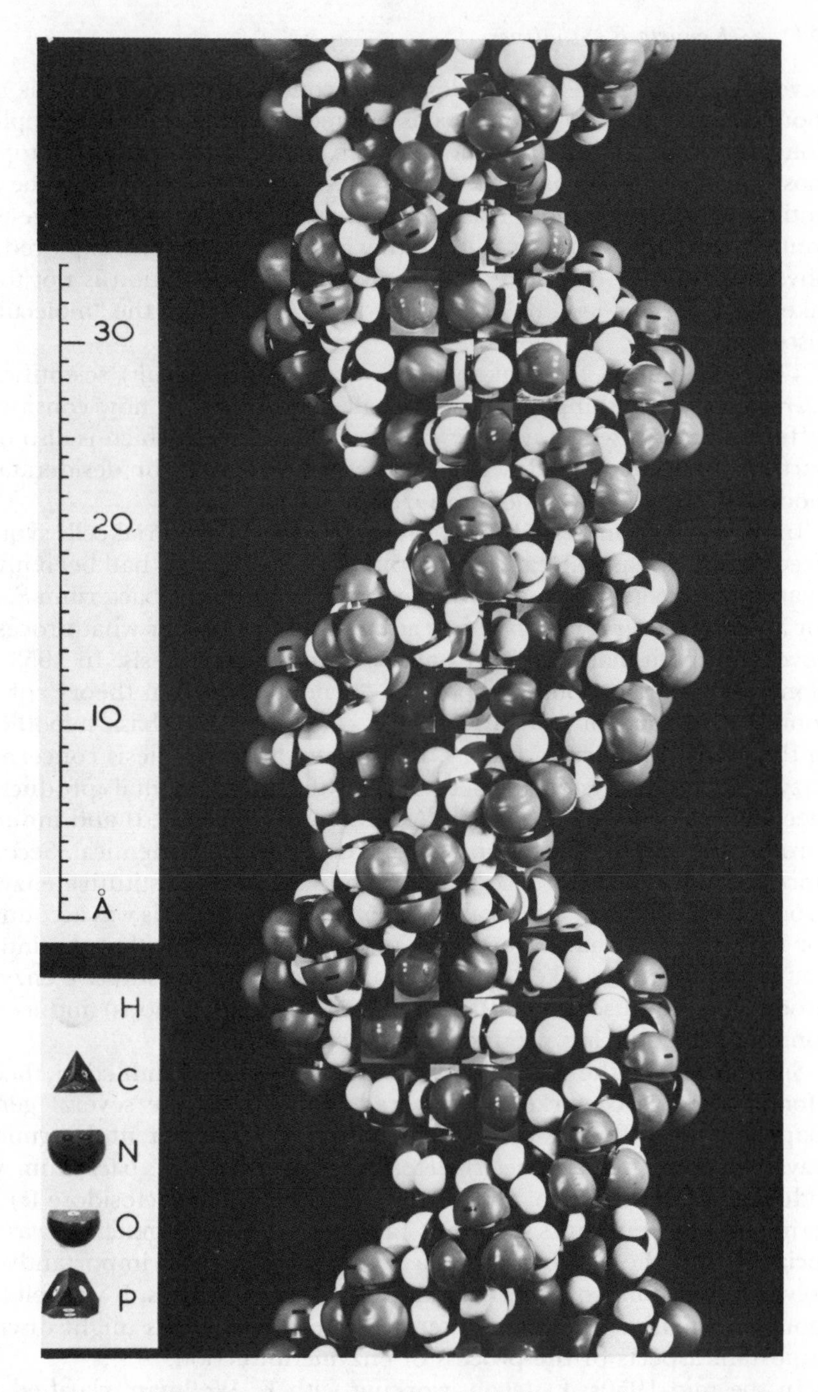

Figure 1. A space-filling model of double-helical DNA. The size of the circles reflects the van der Waals radii of the different atoms. (Courtesy of Professor M. H. F. Wilkins, Department of Biophysics, King's College, London.)

development of molecular biology. A second and related point is that though highly specific problems associated with the uncoiling, replication, and transcription of the double helix are legitimate research topics, most molecular biologists found that this type of problem could be left until later while the more *interesting* issues of protein synthesis, genetic control, and the working out of the genetic code were explored. A physicochemically complete identification and explanation is not to be taken as a *sine qua non* for developing theories utilizing the "molecular" discovery.

The question as to what makes a problem (or result) scientifically *interesting* is something to which I shall return. Let us now consider a different area in which a theory has been developed which is also only partially molecularly defined and in which the goals or desiderata of models or theories become more clearly specified.

In the early to mid-1950s, the process by which bacterial cells synthesized enzymes was unclear. By this time Jacques Monod had been investigating the *lac* genetic region of the common intestinal bacterium *E. coli* for about ten years.[28] Monod was attempting to discover what processes governed the initiation and cessation of enzyme synthesis. In 1953 he, along with Melvin Cohn, proposed a "general induction theory" of the control of enzyme (or protein) synthesis. There were two basic hypotheses in the general induction theory. First, a unitary hypothesis concerning enzyme-forming systems asserted that the fundamental production mechanisms of both constitutive (continuously produced) and induced (production can be turned on and off) enzymes were identical. Second, since there was a difference between induced and constitutive enzyme production (the latter type is always being produced), this was accounted for by supposing that induction is really universal, but that the induction mechanism is masked or hidden in the cases of constitutive enzyme production because a "natural" internal (or endogenous) inducer is continuously inducing production of those enzymes.[29]

Subsequent to the articulation of the generalized induction theory Monod and his coworkers discovered that there were several genes, mappable at different loci, which interacted in as yet undetermined ways. An inducibility gene (*i*) determined whether the bacterium was inducible or constitutive, a gene for the enzyme β-galactosidose (*z*) determined that enzyme's synthesis, and an additional gene (*y*) was associated with the production of a "permease" enzyme importantly involved in the ability of external inducers to enter the cells. It was felt that more information about the interactions of these genes might disclose important aspects of the process of enzyme induction.

In the mid-1950s, F. Jacob, working with E. Wollman, clarified the mechanism of chromosomal transfer from male to female *E. coli* cells.[30] Thus, a most important genetic tool became available for analyzing the interaction of the *i, z,* and *y* genes in the *lac* region of the *E. coli* (see figure 2). An experiment was planned in early 1957, and the specific

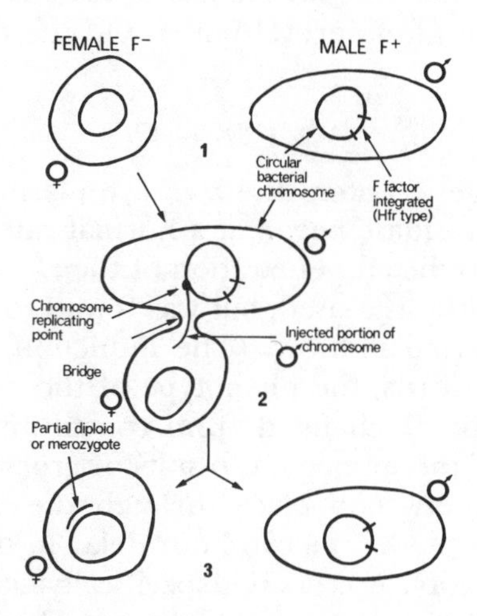

Figure 2. Bacterial conjugation (primitive sexuality). In stage 1, an initially separate male *E. coli* bacterium possessing an integrated *F* or fertility factor (Hfr strain), and a female *E. coli* bacterium (F^-) come into contact (stage 2) and form a protoplasmic bridge between them. The male chromosome begins to replicate and inject the replica into the female bacterium's cytoplasm. The initiation point for replication and insertion is universal for specific strains, and the time of entrance of a particular gene, for example, the *z* gene, is thus predictable. The conjugating bacteria may be separated without damaging them (stage 3) by agitating them in a blender. This results in only a portion of the male chromosome having entered the female; the female is thus a partial diploid.

mutants to be used were selected for in Jacob's laboratory. With the assistance of A. Pardee, who visited the Institut Pasteur in 1957–1958, Jacob and Monod conjugated strains of *E. coli* that were variously inducible and constitutive for the production of β-galactosidase and permease. The details of the experiment, often referred to as the "Pa-ja-mo" or "Pajama" experiment, as regards the types of techniques and experimental controls, do not concern us here. Essentially I wish to focus on two types of crosses or mating that were carried out. (In what follows, I shall observe the convention that i^+ shall stand for the gene determining the inducible phenotype, i^- for the gene determining the constitutive phenotype, z^+ for the normal or wild type gene responsible for the production of β-galactosidase, and z^- for a nonfunctional β-galactosidase gene.)

The result of the cross of bacteria of genotypes:

$$♂ z^-i^- \times ♀ z^+i^+ \tag{I}$$

which feeds the z^-i^- genes into the cytoplasm of the female z^+i^+ type and which produced partially diploid or merozygotic "progeny" with genotype z^-i^-/z^+i^+, was that no enzyme synthesis was detected either immediately or even after several hours. On the other hand, with the cross:

$$♂z^+i^+ \times ♀\, z^-i^- \qquad \text{(II)}$$

in which the z^+i^+ gene enters the z^-i^- cytoplasmic environment, the synthesis of β-galactosidase began at a normal rate about three to four minutes after the predicted penetration of the z^+ gene. After about two hours, however, synthesis ceased, but was found to be restartable by the addition of an external inducer. (The induction kinetics is shown in figure 3.) In other words, the phenotype of the merozygotes resulting from a cross of type II changed from constitutive to inducible. The absolute contrast of the immediate results of crosses I and II indicated that the *i-z* interaction took place through the cytoplasm. This also seemed to be supported by the two-hour delay in the change from constitutivity to inducibility. A cross not specifically cited here showed that no cytoplasmic substances are transferred during conjugation, in agreement with results obtained several years before by Jacob and Wollman, so the genes must have been the cause of the change. It is clear, then, using traditional genetic language, that inducibility is "dominant" over constitutivity, though the dominance is slowly expressed.

According to the generalized induction theory, constitutivity should have been dominant over inducibility. The reversal of this expected

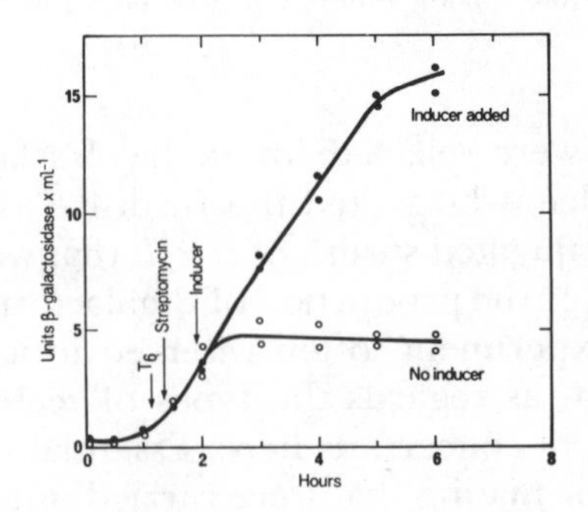

Figure 3. Enzyme kinetic graph representing the results of the Pajama experiment, cross II. The graph clearly shows the commencement of β-galactosidase synthesis very soon after the entrance of the z^+ gene, and also the conversion of the merozygotes from the constitutive form to the inducible form after about two hours. (Reprinted, with the editors' approval, from A. B. Pardee, F. Jacob, and J. Monod, "The Genetic Control and Cytoplasmic Expression of Inducibility in the Synthesis of β-galactosidose of *E. coli*," *Journal of Molecular Biology,* 1 [1958]: 173.)

result by the Pajama experiment seemed to suggest another interpretation of the control of enzyme synthesis. In the preliminary three-page announcement of their results, Pardee, Jacob, and Monod wrote:

> It had appeared reasonable up to the present to suppose that constitutive mutants of a given system synthesized an endogenous inducer absent in the case of the inducible type. The results described here suggest a hypothesis which is exactly opposite. The facts are explained if one supposes that the *i* gene determines (through the mediation of an enzyme) the synthesis, not of an inducer, but of a "repressor" which blocks . . . synthesis of the β-galactosidase, and which the exogenous inducers remove [déplacent] restoring the synthesis. The i^- allele present in the case of the constitutives being inactive, the repressor is not formed, the galactosidase is synthesized, the exogenous inducer is without effect.[31]

In their later and considerably more developed 1959 version of the Pajama paper,[32] Pardee, Jacob, and Monod introduced what they termed the repressor model as follows:

> According to the . . . "repressor" model the activity of the galactosidase-forming system is inhibited in the wild type by a specific "repressor" (probably also involving a galactosidic residue) synthesized under the control of the i^+ gene. The inducer is required only in the wild type as an antagonist of the repressor. In the constitutive (i^-), the repressor is not formed, or is inactive, hence the requirement for an inducer disappears.

This model was contrasted with an "inducer" model, characterized as follows:

> According to . . . the "inducer" model, the activity of the galactosidase-forming system requires the presence of an inducer, both in the constitutive and in the inducible organisms. Such an inducer (a galactoside) is synthesized by both types of organisms. The i^+ gene controls the synthesis of an enzyme which destroys or inactivates the inducer: hence the requirement for an external inducer in the wild type. The i^- mutation inactivates the gene (or its product, the enzyme) allowing accumulation of endogenous inducer.

Both of these models, the authors noted, accounted for the experimental results of the Pajama experiment. Both models also hypothesize the existence of a repressor, which in the repressor model acts directly on the galactosidase enzyme-forming system, and in the inducer model acts indirectly on the natural or endogenous inducer. The two models thus disagree in their conception of the mode of action of the repressor and also disagree over the need for an inducer in enzyme formation.

It is useful to examine the detailed reasons that Pardee, Jacob, and Monod explicitly provided for their choice of the repressor model over the inducer model.[33] These fall into two or three categories: a desire for *unity* of fundamental biological systems, simplicity, and experimental adequacy.

Pardee, Jacob, and Monod first noted that:

> the "repressor" hypothesis might appear strictly ad hoc and arbitrary were it not also suggested by other facts which should be briefly recalled. That the synthesis of certain constitutive enzyme systems may be specifically inhibited by certain products (or even substrates) of their action, was first observed in 1953 by Monod & Cohen-Bazire working with constitutive galactosidase (of *E. coli*) or with tryptophan-synthetase (of *A. aerogenes*) and by Wijesundera & Woods, and Cohn, Cohen & Monod independently working with the methionine-synthetase complex of *E. coli.* It was suggested . . . [then] that this remarkable inhibitory effect could be due to the displacement of an internally-synthesized inducer, responsible for constitutive synthesis, and it was pointed out that such a mechanism could account, in part at least, for the proper adjustment of cellular syntheses. During the past two or three years, several new examples of this effect have been observed and studied in some detail by Vogel, Yates & Pardee, [and] Gorini & Maas. It now appears to be a general rule, for bacteria, that the formation of sequential enzyme systems involved in the synthesis of essential metabolites is inhibited by their end product. The term "repression" was coined by Vogel to distinguish this effect from another, equally general, phenomenon: the control of enzyme activity by end products of metabolism.[34]

The authors added that "the facts which demonstrate the existence and wide occurrence of repression effects justify the basic assumptions of the repressor model. They do not allow a choice between the two models."

It would appear that the type of justification which the existence and wide occurrence of repression effect license is by virtue of what can be characterized as a "principle of the unity of fundamental biological processes." Further reasons that were given to "allow a choice between the two models" that seemed to "make the repressor model seem much more adequate" were associated with simplicity. It was noted that "the repressor model is simpler since it does not require an independent inducer-synthesizing system." This type of simplicity is akin to a generalization of Occam's razor. It should also be noted that this type of simplicity is "inductive" and not simply aesthetic or "descriptive."[35] If an independent inducer-synthesizing system exists in *E. coli,* it should be revealed in certain mutants which had not been encountered. (Such mutants would probably make a form of β-galactosidase with altered tertiary structure.)

A still additional reason offered as a ground for choice between the two models involved experimental adequacy or experimental predictions. An experimental implication of the repressor model is that "constitutive mutants should, as a rule, synthesize more enzyme than induced wild type." Pardee, Jacob, and Monod noted that "this appears to be the case." The authors also referred to a then recent suggestion by F. C. Neidhardt and B. Magasanik concerning the reason for the long-known "glucose effect," whereby the presence of glucose (and other carbohydrates) generally inhibits enzyme synthesis in numerous inducible systems. Neidhardt and Magasanik had suggested that glucose acted as a "preferential metabolic source of internally synthesized repressors." According to the repressor model, Pardee, Jacob, and Monod reasoned, a constitutive mutant should be "largely insensitive to the glucose effect." These authors stated "that this is the case for a very strong argument in favor of the repressor model."

An additional appeal was made to the unity desideratum. Pardee, Jacob, and Monod referred to the consequences of generalizing the inducer model to all biological systems. It is worth quoting them on this matter:

> The inducer model, if generalized, implies that internally synthesized inducers . . . operate in all constitutive systems. This assumption, first suggested as an interpretation of repression effects, has not been vindicated in recent work on repressible biosynthetic systems (Vogel, Gorini and Mass, Yates and Pardee).

Thus, there appears to be an appeal to a unitary hypothesis for all enzyme control systems, and it would seem that such an appeal is ultimately based on what Cohn and Monod had termed in their 1953 paper "the biological faith in the unity and ultimate simplicity of organization of living beings."[36]

Still another aspect of this "unity of biological systems" was appealed to by Pardee, Jacob, and Monod at the very end of their paper, though at that point the unity appeared under the guise of a rather striking formal analogy between the results of the Pajama experiment and Jacob's and Wollman's work on the regulation of immunity to, and zygotic induction of, certain forms of bacteriophage which attacked *E. coli.* As Pardee, Jacob, and Monod noted, "the formal analogy between this situation [the Pajama experiment as interpreted by the repressor model] and that which is known to exist in the control of immunity and zygotic induction of temperate bacteriophage is so complete as to suggest that the basic mechanism might be essentially the same."

Two general philosophical points need to be underscored at the conclusion of this long historical analysis. First, the chemically incomplete aspects of the models, either the Watson-Crick model mentioned earlier or the repressor model as developed above, should be kept in mind. It is a contention of the present essay that the explanation for tolerating such chemical incompleteness is that reduction in the directional sense is a goal of significantly lesser priority compared to other desiderata of a research program. Some of these other desiderata were earlier introduced as criteria of theory choice, namely, unity, simplicity, and empirical adequacy. These criteria are partly constitutive of, and partly signs of, the intelligibilistic and pragmatic goals of scientific inquiry. A *reductionistic* research strategy *may* yield a theory (or model), or a modification of a theory, which significantly increases the empirical adequacy of the theory or, say, results in a simplification of it, but it *need* not. The strategy which is expected to maximize these goals or values of science is usually the strategy selected. Furthermore, it would appear that problem solutions which have the properties of unity and so forth are those solutions which are "interesting" to scientists.

I shall not in these pages attempt to analyze in any detail the relations between what might be considered more primary values of science such as understanding, prediction, and control, and the possibly more derivative or secondary values such as experimental adequacy, unity, and simplicity. In my opinion these latter values are constitutive of what we mean by understanding, and they aid prediction and control. Theories which organize a domain of experimental inquiry in simple ways and which explain (and predict) large classes of experimental results are desirable on the basis of both primary and secondary values, as characterized immediately above.

It is also important to realize that as research goes forward, additional experimental facts may be disclosed which alter the amount of information on various levels. Specifically, a new genetically definable mutation may be encountered, for example, a promoter locus mutation in the operon theory. Alternatively, an investigator might, on the molecular level, discover a new sequence of nucleotides common to a large variety of translation phenomena, for example, a punctuation codon. These new facts may cause modification of the theories which could enter into a reduction relation, and it might well be a lower research priority to reformulate the connectability assumptions or reduction functions in a molecular direction than to develop a more general theory at a nonmolecular level. Therefore, both scientific progress *and* a focus on more important research goals may (and often do, according to my reading of a number of historical episodes in the biomedical sciences) direct scientific activity away from directional reductionistic research programs.

The orientation of science toward powerful unified explanatory structures which explain and afford means of intervening in nature indicates that there are *values* present *in* science. This goes counter to the claim that science is *per se* "value-free." The issue of the value-free character of science has received scrutiny by a number of distinguished thinkers, and we now turn to a discussion of this subject. This will not only enable us to indicate more specifically the nature of values in science, but will also allow a response to the broader antireductionistic claims that Roszak and Cassell introduced earlier.

4. THE VALUE CONTEXT OF SCIENTIFIC PROGRESS

In the last section I argued that the incomplete and dynamic character of scientific inquiry requires *choices* among hypotheses, theories, research projects, and research programs.[37] These choices, I have contended, are governed by *values* that are intrinsic to science. The existence of these values raises at least two important issues:

1. Whether these values are simply *descriptive* of how scientists *do* choose or whether they are also normative and tell a scientist how he *should* choose, and
2. The much debated question of the value-free nature of science.

Let us look at the second issue first.

It should be clear that on the construal of science presented in section 3 science is *not* value-free. An alternative methodological view of science in which science is understood as a reporting of what exists in a simpler and more economical form—a Mach-like positivistic construal[38]—might have licensed a value-free analysis. Understanding science as a series of hypothetical guesses, some of which are better than others on generally accepted grounds, obviates this approach.

Such a claim is not completely novel. On somewhat narrower grounds, R. Rudner argued:

> Now I take it that no analysis of what constitutes the method of science would be satisfactory unless it comprised some assertion to the effect that the scientist as scientist accepts or rejects hypotheses.
>
> But if this is so, then clearly the scientist as scientist does make value judgments. For, since no scientific hypothesis is ever completely verified, in accepting a hypothesis the scientist must make the decision that the evidence is "sufficiently" strong or that the probability is "sufficiently" high to warrant the acceptance of the hypothesis. Obviously our decision regarding the evidence and respecting how strong is "strong enough" is going to be a function of the "importance," in the typically ethical sense, of making a mistake in accepting or rejecting the hypothesis.[39]

Rudner's claim was subsequently questioned by R. Jeffrey and I. Levi. Jeffrey proposed that "the activity proper to the scientist is the assignment of probabilities (with respect to currently available evidence) to the hypotheses, which on the usual view, he simply accepts or rejects."[40] Levi contended that scientists can still accept or reject hypotheses and accept a thesis of the value-neutrality of science.[41] The debate was more recently joined by J. Leach, who argued that scientists do make value judgments in accepting or rejecting hypotheses, though these are of a "pragmatic" sort.[42]

It would take us beyond the scope of the present essay to analyze this interchange, and it is mentioned primarily to indicate that the issue has been considered in some depth in the earlier literature. A later claim that science involved a series of value choices was made by N. Rescher, who argued that, contrary to widespread belief in the value-neutral character of science, the discipline could be shown to contain value elements in the choice of research goals, the selection of research methods, the specification of standards of proof, and the allocation of credit for research accomplishments, among others.[43]

A more interesting issue than the value-free character of science involves the nature of the values. This was mentioned as point (1), in which the normative-versus-descriptive issue was raised. The problem can be posed as follows: Suppose we are able to demonstrate by an examination of various scientific documents or interviews with scientists that there is a widespread common value structure endemic in scientific inquiry. A critic might well raise the question of whether these values have any normative force.[44] David Hume long ago pointed out the logical problem of deriving an "ought" (or normative) statement from an "is" (or descriptive) statement.

This question of the normative character, then, goes to the heart of the value issue in science, for it is these normative kinds of values in which there is most interest. The question also discloses that the issue of the value-free character of science is closely connected with one of the more disputed problems in philosophical ethics: the normative-descriptive distinction.

It is not possible here to rehearse all of the suggestions philosophers have offered to bridge the descriptive-normative divide. I will propose one possibility which I believe both does justice to discerned interactions between the history of science and the philosophy of science, and can be utilized in the broader context of examination of progress in the biomedical sciences, and which represents in a programmatic way the direction in which a broadly "reductionistic" theory of values should be developed.

Basically my analysis of the interaction between the normative and the descriptive in science comes from the naturalistic, humanistic, and pragmatic schools of philosophical inquiry, and is indebted to suggestions made in various places by C. S. Peirce, John Dewey, and C. I. Lewis.[45] Let me first outline the general position very briefly. I shall then develop it in more detail and apply the theory of value to the issues of the present essay.

We can, following Dewey, distinguish between "prizings" and "appraisings." (The words are not important, though the concepts to which they refer are.) "Prizings" are those aspects of our experience toward which we gravitate and aim prior to critical and reflective thought; they are intuitively perceived as values. "Appraisings," on the other hand, are likings which are the outcome of a multilevel complex of prizings subjected to conscious scrutiny in which the conditions and consequences of possible actions are considered. The "good," that at which one *should* aim, is *constructed:* It is *defined* as "good." This process of *intelligent* choice, it can be argued, yields maximal individual and social *satisfaction.* The ethical theory is thus eudaemonistic. The approach considers given values as continually subject to scrutiny. It is a public approach in which arguments about the desirability of values and conflicts between incompatible values can be entertained.

Let me now attempt to develop and defend this position in some detail, and also to anticipate some objections. This elaboration will primarily draw from a number of Dewey's ethical writings, but will, I think, be congruent with similar pragmatic positions developed by C. S. Peirce and C. I. Lewis.

It is most appropriate to begin an elaboration of a pragmatic theory of valuation by noting that it ultimately rests on an *intuitionistic* basis. Contrary to Lewis's intuitionistic position, however, the intuitions which one has of values are for Dewey—I think correctly—both *fallible* and *corrigible,* and also what I will term *multilevel.*

For Dewey, intuitions of value are "tertiary qualities" of nature, a term he borrowed from Santayana. In a defense of his moral theory against H. W. Stuart,[46] Dewey wrote:

> Instead of presenting that kind of mechanistic naturalism that is bound to deny the "reality" of the qualities which are the raw material of the values with which morals is concerned, I have repeatedly insisted that one theory of Nature be framed on the basis of giving full credence to these qualities just as they present themselves.[47]

Common elements of these intuitions can be abstracted into what Dewey, in his *Theory of the Moral Life,*[48] termed *principles.* These principles can also be given at the *beginning* of an ethical situation, being embodied in traditional moral codes, and in "legal history, judicial decisions, and legislative activity," as well as in the various sciences such as medicine, sociology, and economics, and also in rational philosophical theories of ethics.[49]

Taken as starting points, these principles can themselves be both prized and appraised in the light of other prizings and consequences. It is only if these principles take on the status of what Dewey called *rules,* that is, inflexible and incorrigible principles, that this form of intuitionism degenerates into traditional intuitionism,[50] an ethical theory which Dewey saw as lending support to an ossified conservatism.

We have, then, intuitions on the level of individual likings, and also on the level of principles. Conflicts and assessments are to be mediated by the experimental method, a method which Dewey contended is better than alternative methods such as authority, tradition, rationalism, or revelation. The experimental method is the action of intelligence such as we find employed in science. It would seem natural to conclude on the basis of the pragmatic theory of valuation as developed thus far that the *justification for the experimental method* is based on the same corrigible intuitionism which is encountered operating in the levels of concrete experience and principles.[51] If this reconstruction of Dewey's view is defensible, we are presented with a theory of valuation which is based on a multilevel and corrigible intuitionism. These "levels" are themselves almost certainly artificial divisions on the continuum of generality, certain aspects of which are at any one time presented with force, others of which are more remote. As time and experience change, different aspects of experience come more to the fore and others recede. These multilevel intuitions are thus always partial and may occasionally be flagrantly wrong, but they are not so unstable that an accretion of experience, which often passes for reflective and sensitive wisdom, is not the outcome of the process.[52]

It is this view which I think is embodied in Dewey's pragmatic theory of value, a theory he summarized somewhat cryptically in his *Theory of Valuation* in the following manner:

> These considerations lead to the central question: What are the conditions that have to be met so that knowledge of past and existing valuations becomes an instrumentality of valuation in formation of new desires and interests—of desires and interests that the test of experience show to be best worth fostering? It is clear upon our view that no abstract theory of valuation can be put side by side, so to speak, with existing valuations as the standard for judging them.
>
> The answer is that improved valuation must grow out of existing valuations, subjected to critical methods of investigation that bring them into systematic relations with one another. Admitting that these valuations are largely and probably, in the main, defective, it might at first sight seem as if the idea that improvement would spring from bringing them into connection with one another is like recommending that one lift himself by his bootstraps. But such an impression arises only because of failure to consider how they actually may be brought into relation with one another, namely, by examination of their respective conditions and consequences. Only by following this path will they be reduced to such homogeneous terms that they are comparable with one another.
>
> This method, in fact, simply carries over to human or social phenomena the methods that have proved successful in dealing with the subject matter of physics and chemistry. (p. 60)

The method here outlined may receive some additional support from the fact that it also appears to be the method employed by a variety of nonpragmatic theorists. For example, John Rawls articulates his method for arriving at the generalizations which are to be taken as the basis for his deontological contract theory construal of ethics as follows:

> In searching for the most favored description of this situation [the "original position" in which rational beings behind a "veil of ignorance" choose the ethical principles which will bind them] we work from both ends. We begin by describing it so that it represents generally shared and preferably weak conditions. We then see if these conditions are strong enough to yield a significant set of principles. If not, we look for further premises equally reasonable. But if so, and these principles match our considered convictions of justice, then so far well and good. But presumably there will be discrepancies. In this case we have a choice. We can either modify judgments, for even the judgments we take provisionally as fixed points are liable to revision. By going back and forth, sometimes altering the

conditions of the contractual circumstances, at others withdrawing our judgments and conforming them to principle, I assume that eventually we shall find a description of the initial situation that both expresses reasonable conditions and yields principles which match our considered judgments duly pruned and adjusted. This state of affairs I refer to as *reflective equilibrium.*[53]

Something like what Rawls attempts in the remainder of his book is, of course, a necessary further step in the articulation of a complete system of values. It is not my intention in these pages to present anything like a *complete* system but rather to suggest the *method* of valuation which allows for a proper perception of the value aspects of scientific inquiry.

Within the context of the history of science, this pragmatic theory of values or valuation allows one to move from (relatively) descriptive values found in historical episodes ("relatively" because though these values are based on intuitions, the complex of intuitions has not yet been adjudicated and clarified) to (relatively) normative, more philosophically sophisticated values. These would be the products of articulation, scrutiny, and debate about the aims of science in the light of detailed cases. For example, "simplicity" as a natural desideratum of a scientist can be scrutinized in the light of other goals; and from the point of view of a desirable, more precise explication, it can be tested in a series of case studies, and a reasoned normative conclusion can be reached. This process seems to accord with scientific practice and also allows for an *evolution* of the desiderata of science as a function of evolving science.[54]

Having now proposed the pragmatic theory of valuation as the most appropriate approach to understanding the nature and role of values in science (and outside science), let us return to some of the issues of reductionism in the biomedical sciences.

I argued in section 3 that directional reductionism as an aim of a research project or program was often of considerably lower priority than planning further experiments to understand a biological mechanism on a higher, say genetic or cellular, level. In reconstructing actual historical moves, however, one also has to take into account types of constraints, additional to those mentioned earlier, which are primarily "internal." We will understand as "internal" considerations those elements of scientific decision-making having to do with the stage of maturity of a theory, the sophistication and precision of experimental techniques, and various criteria of choice affecting these elements, such as relative simplicity. We will term "external" considerations those elements such as political proscription, societal needs, a religiously engendered ethics, economic developments, patterns of funding by various foundations and agencies, and the peer pressure which sees some scientific areas as more exciting than others. The last of these clearly is at least in part a function of "internal" considerations, which brings me to my point. To provide an adequate historical reconstruction of the process of scientific change, all relevant explanatory considerations need to be invoked, some of which fall into that group which is traditionally termed "external," some "internal," and some interdependent on these two classes. Attention to the broad spectrum of desiderata and values on individual, institutional, and societal fronts should provide the clearest possible rational reconstruction of the progress of various research projects and programs.[55] The reasons, then, for the reductionistic or nonreductionistic direction of different investigations will become obvious only in the context of this broad framework.

The incorporation of values "external" to science into an account of scientific change suggests that we consider the question of scientific advance in the biomedical sciences from this broader perspective. In the context of this perspective I shall argue, contrary to what the late distinguished molecular biologist Jacques Monod recently proposed, that a strictly internal scientific ethic would not be acceptable. Further, however, in contrast to Roszak's and Cassell's suggestions discussed in section 1, I would argue that *scientific thinking and methods* supplemented with extrascientific "prizings" afford an appropriate ethic. Let us look at this issue specifically in the microcosm of the biomedical sciences, where the issues become more obvious.

Let us first consider what a strictly "internal" ethic of science might be like, even if only in the broadest of outlines. In his *Chance and Necessity,* J. Monod proposed an "ethic of knowledge" with the following properties:

> True knowledge is ignorant of values, but it cannot be grounded elsewhere than upon a value judgment, or rather upon an *axiomatic* value. It is obvious that the positing of the principle of objectivity as the condition of true knowledge *constitutes an ethical choice and not a judgment arrived at from knowledge, since, according to the postulate's own terms, there cannot have been any "true" knowledge prior to this arbitrary choice.* In order to establish the *norm* for knowledge, the objectivity principle defines a *value:* that value is objective knowledge itself. Thus, assenting to the principle of objectivity one announces one's adherence to the basic statement of an ethical system, one asserts the ethic of knowledge. . . .
>
> The ethic of knowledge that created the modern world is the only ethic compatible with it, the only one capable, once understood and accepted, of guiding its evolution. . . .
>
> Where then shall we find the source of truth and the moral inspiration for a really scientific socialist humanism, if not in the sources of science itself, in the ethic upon which knowledge is founded, and which by free choice *makes knowledge the supreme value*—the measure and warrant for all other values? An ethic which bases moral responsibility upon the very freedom of that axiomatic choice. Accepted as the foundation for social and political institutions, hence as the measure of their authenticity, their value, only the ethic of knowledge could lead to socialism.[56]

Now I submit that such an ethic, *without additional constraints and/or values,* which is devoted to maximizing scientific knowledge (since science is, according to Monod, the acme of objective knowledge), would have undesirable consequences. For example such an ethic *could* be construed as legitimating forced experimentation on human beings such as was practiced by the Nazis in World War II. Human experimentation is necessary for full knowledge of the human biological system because of species-specific differences. It is unclear that some (perhaps low-IQ) individuals would not be forced, under an ethic of knowledge, to be sources of important knowledge for the majority. Additional safeguards appear to be necessary, and I submit that they are available in the general ethical codes, constitutions, and philosophical statements which urge that each human being be considered an end in himself or herself. Such an ethical approach still permits the use of humans in biomedical experimentation, but only on the basis of their voluntary "informed consent." Taking the autonomy of each human being as one of our highest values is not entailed by a strictly "internalistic" ethic (or at least is not entailed without further premises or additional argument). If we broaden the initial field of an evaluation to encompass both values intrinsic to science *and* values that our humanistic heritage has judged to be necessary conditions for a good life and a just society, we do not require such an (at least questionable) entailment. We then would have a *pluralistic* set of values in which the external, humanistically given values set appropriate boundaries within which the scientific values can usefully function to move science forward.

A continued reevaluation of our goals on both scientific and societal fronts is suggested by the pragmatic approach articulated earlier. It is important to note that the pragmatic approach utilizes all of the *methods* science can offer *as applied to* a broad pluralistic field of initial values to yield a set of values that are legitimated by critical analytical reflection. If this latter species of thinking is construed as reductionistic, then I have accomplished my initial program in which I proposed that it could be shown how a reductionist, in all senses, can nonetheless champion research on many different biological levels and also engage in value-conditioned and value-laden thinking in science.

NOTES

I would like to thank David Hull and Allan Gibbard for comments on an earlier version of this essay. I am also indebted to the John Simon Guggenheim Memorial Foundation and to the National Science Foundation for support of research.

1. For a construal of reductionism as a methodological approach to biology, see the paper by P. Medawar in *Studies in the Philosophy of Biology,* ed. F. Ayala and T. Dobzhansky (London: Macmillan, 1974).

2. See K. Schaffner, "Antireductionism and Molecular Biology," *Science,* 157 (1967):644–47.

3. T. Roszak, *Where the Wasteland Ends* (New York: Doubleday, 1972), p. 263. Roszak's views were taken seriously by the influential publication *Science.* See N. Wade, "Theodore Roszak: Visionary Critic of Science," *Science,* 178 (1972):960–62.

4. E. J. Cassell, "Preliminary Explorations of Thinking in Medicine," *Ethics in Science and Medicine,* 2 (1975):1–12.

5. T. Dobzhansky, "On Some Fundamental Concepts of Darwinian Biology," in *Evolutionary Biology,* 2, ed. T. Dobzhansky, M. Hecht, and W. Steere (New York: Appleton-Century-Crofts, 1968), esp. pp. 1–2.

6. K. Schaffner, "The Peripherality of Reductionism in the Development of Molecular Biology," *Journal of the History of Biology,* 7 (1974):111–39.

7. Analyses of human values in the biomedical sciences in connection with experiments on human subjects have proliferated in recent years. J. Katz's anthology, *Experimentation with Human Beings* (New York: Russell Sage, 1972), edited with the assistance of A. Capron and E. Glass, is an extensive compilation of documents relating to human experimentation. B. Barber, J. Lally, J. Makarushka, and D. Sullivan's *Research on Human Subjects* (New York: Russell Sage, 1973), is an important study of peer-review practices and socialization of ethical values in medical training. It has recently been supplemented with B. Gray's *Human Subjects in Medical Experimentation* (New York: Wiley Interscience, 1975), a close analysis of human subjects engaged in medical research projects. C. Fried has provided trenchant legal-ethical analyses of human experimentation and the randomized clinical trial in his *Medical Experimentation* (New York: American Elsevier, 1974). A two-day forum sponsored by the National Academy of Sciences in 1975 also examined these issues in *Experiments and Research with Humans: Values in Conflict* (Washington, D.C.: National Academy of Sciences, 1975).

8. E. Nagel, "Reduction in the Natural Sciences," in *Science and Civilization,* ed. R. Stauffer (Madison: University of Wisconsin Press, 1949), pp. 99–145.

9. E. Nagel, *The Structure of Science* (New York: Harcourt, Brace, and World, 1961).

10. Ibid., pp. 353–54.

11. Ibid., pp. 354–58.

12. See H. Feigl, "The Mental and the Physical," in *Minnesota Studies in the Philosophy of Science,* 2, ed. H. Feigl, M. Scriven, and G. Maxwell (Minneapolis: University of Minnesota Press, 1958), pp. 370–497; and L. Sklar, "Intertheoretic Reduction in the Natural Sciences" (Ph.D. dissertation, Princeton University, 1964), for accounts of synthetic identities in reduction. Also see K. Schaffner, "Approaches to Reduction," *Philosophy of Science,* 34 (1967):137–47; K. Schaffner "The Watson-Crick Model and Reductionism," *British Journal for the Philosophy of Science,* 20 (1969):325–48; and R. Causey "Attribute-Identities in Microreductions," *Journal of Philosophy,* 69 (1972):407–22.

13. P. K. Feyerabend, "Explanation, Reduction, and Empiricism," in *Minnesota Studies in the Philosophy of Science,* 3, ed. H. Feigl and G. Maxwell (Minneapolis: University of Minnesota Press, 1962), pp. 28–97.

14. See Watkins's personal communication, cited in Feyerabend, "Explanation," p. 93, and Schaffner, "Approaches to Reduction."

15. See Schaffner, "Approaches to Reduction"; "Peripherality of Reductionism"; and "Reductionism in Biology: Prospects and Problems," in *PSA-1974,* ed. A. Michalos and R. S. Cohen (Dordrecht, Holland: Reidel, 1976).

16. See K. Schaffner, "Correspondence Rules," *Philosophy of Science,* 36 (1969):280–90; and idem, "Outlines of a Logic of Comparative Theory Evaluation with Special Attention to Pre- and Post-Relativistic Electrodynamics," in *Minnesota Studies in the Philosophy of Science,* 5, ed. R. Stuewer (Minneapolis: University of Minnesota Press, 1970), pp. 311–53, 365–76, esp. 315–16.

17. D. Shapere, "Scientific Theories and Their Domains," in *The Structure of Scientific Theories,* ed. F. Suppe (Urbana: University of Illinois Press, 1974), pp. 518–565.

18. Schaffner, "Approaches to Reduction," "Peripherality of Reductionism," and "Reductionism in Biology."

19. See M. Ruse, "Two Biological Revolutions" and "Reduction, Replacement, and Molecular Biology," *Dialectica,* 25 (1971):17–72; and D. Hull, *Philosophy of Biological Science* (Englewood Cliffs, N.J.: Prentice-Hall, 1974), esp. chap. 1.

20. See Schaffner, "Peripherality of Reductionism" and "Reductionism in Biology."

21. See Schaffner, "Correspondence Rules" and "Outlines of a Logic of Comparative Theory Evaluation."

22. See L. Clendening, ed., *Source Book of Medical History* (New York: Dover, 1942), pp. 39–51.

23. See E. T. Whittaker, *A History of the Theories of Aether and Electricity,* vol. 1, *The Classical Theories* (New York: Harper and Bros., 1960), esp. chaps. 1 and 4; and K. Schaffner, *Nineteenth-Century Aether Theories* (Oxford: Pergamon Press, 1972), esp. chaps. 1 and 2.

24. See Schaffner, *Nineteenth-Century Aether Theories,* chap. 6.

25. See Schaffner, "The Watson-Crick Model and Reductionism," for comments on this issue.

26. J. D. Watson and F. H. C. Crick, "A Structure for Deoxyribose Nucleic Acids," *Nature,* 171 (1953):737–38; and J. D. Watson, "Genetical Implications of the Structure of Deoxyribose Nucleic Acid," *Nature,* 171 (1953):964–69.

27. The uncertainty involved in the biochemical basis of DNA replication is discussed with references to the literature in "Unwinding the Double Helix," *Nature,* 227 (1970): 1294. Two alternative and formally inconsistent models for the transcription of DNA into RNA were provided by V. L. Florentiev and V. I. Ivanov, "RNA Polymerase: Two-Step Mechanism with Overlapping Steps," and by P. A. Riley, "A Suggested Mechanism for DNA Transcription," *Nature,* 228 (1970):519–22 and 522–25. Problems with DNA unwinding and a discussion of the different DNA polymerases are to be found in "Does DNA Unwind?" and "Kornberg Repairs Enzyme," *Nature New Biology,* 231 (1971):193–94.

28. The account that follows is essentially based on my analysis of the genesis of the repressor hypothesis presented in my "Logic of Discovery and Justification in Regulatory Genetics," *Studies in History and Philosophy of Science,* 4 (1974):349–85.

29. See M. Cohn and J. Monod, "Specific Inhibition and Induction of Enzyme Biosynthesis," in *Adaptation in Microorganisms,* ed. R. Davies and E. F. Gale (Cambridge: Cambridge University Press, 1953), pp. 132–49.

30. See F. Jacob, "Genetics of the Bacterial Cell," *Science,* 152 (1966):1470–79, for history and references.

31. A. B. Pardee, F. Jacob, and J. Monod, "Sur l'expression et le rôle des allèles "inductible" et "constitutif" dans la β-galactosidose chez des zygotes d'*Escherichia coli,*" *Comptes Rendus des Séances de l'Académie des Sciences,* 246 (1958):3125–26. My translation.

32. A. B. Pardee, F. Jacob, and J. Monod, "The Genetic Control and Cytoplasmic Expression of Inducibility in the Synthesis of β-galactosidose of *E. Coli,*" *Journal of Molecular Biology,* 1 (1958):175.

33. There may be (and I would argue for) additional classes of reasons for accepting one hypothesis or theory over its competitor(s) than I present here. One important class of reasons not explicitly cited by Pardee, Jacob, and Monod is what I have termed elsewhere "theoretical context sufficiency." This refers to the desire of scientists not to accept a new theory which conflicts with other, well-established *theories.*

34. See Pardee, Jacob, and Monod, "Genetic Control and Cytoplasmic Expression," p. 176. References to other papers cited by Pardee, Jacob, and Monod have been deleted. See their "Sur l'expression et le role des alleles" or Schaffner, "Logic of Discovery and Justification," for complete references.

35. See H. Reichenbach, *Experience and Prediction* (Chicago: University of Chicago Press, 1938), for an analysis of the distinction between "descriptive" and "inductive" simplicity.

36. Cohn and Monod, "Specific Inhibition and Induction of Enzyme Biosynthesis."

37. Scientists choose among various alternatives. I have listed the choices in accordance with different magnitudes of reorientation of inquiry. A hypothesis is one of the smaller modifications; acceptance of a new theory represents a more serious (and more interesting) choice. Choices can occur among relatively small units which I shall term "research projects," consisting of one (or several) problem(s), various previous experimental results, and a complex of theories, some well established and some programmatic. The research project is one of many which can be undertaken in the context of the same "research programme"—a term which has roughly the same sense in which Lakatos introduced it as, I believe, a less mystical entity than a Kuhnian "paradigm." See I. Lakatos, "Falsification and the Methodology of Research Programmes," in *Criticism and the Growth of Knowledge*, ed. I. Lakatos and A. Musgrave (Cambridge: Cambridge University Press, 1970), pp. 91–195.

38. See E. Mach, *Analysis of Sensations*, 5th ed. (1906; reprint ed., New York: Dover, 1959).

39. R. Rudner, "The Scientist qua Scientist Makes Value Judgements," *Philosophy of Science*, 20 (1953):1–6.

40. R. C. Jeffrey, "Valuation and Acceptance of Scientific Hypotheses," *Philosophy of Science*, 23 (1956):237.

41. I. Levi, "Must the Scientist Make Value Judgements?", *Journal of Philosophy*, 57 (1960):345–57.

42. J. Leach, "Explanation and Value Neutrality," *British Journal for the Philosophy of Science*, 19 (1968):93–108.

43. N. Rescher, "The Ethical Dimension of Scientific Research," in *Beyond the Edge of Certainty*, University of Pittsburgh Series in the Philosophy of Science, 2, ed. R. Colodny (Englewood Cliffs, N.J.: Prentice-Hall, 1965), pp. 261–76. Also see M. Scriven, "The Exact Role of Value Judgements in Science," in *PSA-1972*, ed. K. F. Schaffner and R. S. Cohen (Dordrecht, Holland: Reidel, 1974) pp. 219–47, for a still more recent criticism of the value-free nature of science.

44. For example, see R. Giere, "History and Philosophy of Science: Intimate Relationship or Marriage of Convenience," *British Journal for the Philosophy of Science*, 24 (1973):282–97.

45. See C. S. Peirce, *Collected Papers*, ed. C. Hartshorne and P. Weiss (Cambridge, Mass.: Harvard University Press, 1965), vol. 1, pp. 314–21, and vol. 5, pp. 24–28, 77–93; J. Dewey, *Theory of Valuation* (Chicago: University of Chicago Press, 1939) and "The Construction of Good," in his *Quest for Certainty* (New York: Putnam, 1960), chap. 10 (first published in 1929); and C. I. Lewis, *An Analysis of Knowledge and Valuation* (La Salle, Ill.: Open Court, 1946).

46. H. W. Stuart, "Dewey's Ethical Theory," in *The Philosophy of John Dewey*, 2d ed., ed. P. A. Schilpp (New York: Tudor, 1951), pp. 291–333.

47. J. Dewey, "Experience, Knowledge, and Value: A Rejoinder," ibid., p. 580.

48. J. Dewey, *Theory of the Moral Life* (New York: Holt, Rinehart, and Winston, 1960). (This is a reprint of part II of Dewey and Tufts' 1932 revised edition of their *Ethics*. Page references that follow are to the 1960 reprint.)

49. Ibid, pp. 22–25.

50. Ibid., pp. 136–46.

51. The justification of method by intuition has had some distinguished adherents. C. S. Peirce seemed to accept such a thesis when he admitted that after long reflection he began to be impressed with the dependence of logic on ethics, and both on aesthetics. See his *Collected Papers*, ed. C. Hartshorne and P. Weiss, vol. 2, pp. 114–17. More recently the late

R. Carnap surprised a number of his more positivistically inclined colleagues with his intuitionistic defense of inductive logic. In his "Inductive Logic and Inductive Intuition," in *The Problem of Inductive Logic,* ed. I. Lakatos (Amsterdam: North-Holland, 1968), pp. 258–67, Carnap wrote:

> In order to learn inductive reasoning, we must have what I call *inductive intuition.* My friends warned me against the use of the word 'intuition'. So I hasten to add that I do not mean by 'intuition', here 'inductive intuition', a source of knowledge that is infallible. I don't think there is anything infallible in human knowledge or activity. Certainly inductive intuition may on occasion lead us astray; the same holds for all other kinds of intuition, for example, spatial intuition, which we use in geometry. (p. 265)

52. If this reconstruction of Dewey as a multilevel and corrigible intuitionist is correct, I believe it shows that M. White's influential criticism of Dewey (see his "Value and Obligation in Dewey and Lewis," *Philosophical Review,* 58 (1949):321–29) misses the point, inasmuch as the source of the moral "able" in desirable is given in a complex and changing set of multilevel intuitions. Such a view as I am arguing for here also obviates the need to search for a "practical," as opposed to theoretical, resolution to this problem such as S. Hook presented. (See "The Ethical Theory of John Dewey," in S. Hook, *The Quest for Being* [New York: Delta, 1963], pp. 49–70, esp. pp. 37–38.)

53. J. Rawls, *A Theory of Justice* (Cambridge, Mass.: Harvard University Press, 1971), p. 20.

54. See S. Toulmin, *Human Understanding* (Princeton: Princeton University Press, 1972), and D. Shapere, "Scientific Theories and Their Domains."

55. See Toulmin, *Human Understanding,* pp. 206–22 and 300–08, for a similar thesis.

56. J. Monod, *Chance and Necessity* (New York: Knopf, 1970), pp. 140–42; last italics added.

HADLEY CANTRIL
1906–1969

Psychology and Scientific Inquiry

> It is no use, indeed only dangerous, in the present state of our knowledge with regard to psychology and the physics of the brain, to fill the void of ignorance by hypotheses which can neither be proven nor refuted.
>
> —Karl Pearson
> *The Grammar of Science*

The ultimate goal of psychology is the understanding of human living so that individuals can live more effectively. The psychologist's aim is that of formulating a set of constructs which will enable him conceptually to "understand," "explain," and "predict" the activities and experiences of the functional union we call a behaving person. A prerequisite for psychological research is an understanding of what an individual is aware of from his unique behavioral center in any occasion of living. The word "center" is used here in the dictionary sense of "the point toward which any force, feeling or action tends or from which any force, or influence takes its origin; as a storm center" (*Webster's Collegiate,* 5th ed.).

The psychologist interested in understanding the process of living must start from naive experience in the phenomenal area, for only then will he be able to undertake investigations that will disclose the nature of the processes playing a role in behavior of which the experiencing individual is taking account. The words "experience" and "behavior" are used here as *interdependent* abstractions man has created, neither of which would be meaningful except for the other.

As anyone knows, the task of starting with naive experience and formulating systematic constructs on the basis of such experience is a particularly difficult undertaking. The very essence of living is its unity and flow. Yet in order to get any grasp of his most complicated subject matter and to communicate his understanding to others, the psychologist is forced to break up into distinguishable parts what really constitutes an indivisible, functional aggregate; he is forced to consider separately the various aspects of living which are all interrelated and interdependent and none of which would function as they do except for all the others. The various aspects of living that are experienced are never separate from each other but fuse together to form this living. The psychologist must try to orchestrate into a single symphony the harmonious or discordant sounds with motifs, themes, phrases, accents, and repetitions related to the separate notes played by the many instruments.

If the psychologist, then, is to be faithful to his subject matter, he must always bear in mind that living is an orchestration of ongoing processes and that any process he may choose to differentiate out of this aggregate can be understood only if it is recognized as referring to a phase of man's orchestrated functioning. It is, for example, a commonplace of philosophical and psychological thinking that "cognitive" and "motor" processes are themselves distinctions that can be misleading unless there is full cognizance of the fact that there can be no "knowing" without "doing," just as there can be no "person" except for an "environment," nothing "subjective" except for what is "objective," nothing "personal" except for what is "social," nothing experienced as existing "out there in its own right" except for the organizations and significances to an individual of the happenings going on in the world around him which he associated with light waves, sound waves, and so forth as instruments of explanation.

A necessary part of our consideration is, of course, that the conceptual abstractions we use are several steps removed from our primary data, namely, naive experience. In order to bridge the gap between naive experience and conceptual abstractions, the psychologist must consider areas of complexity and abstraction that become progressively further removed from his first-order data. It is imperative that any investigator be aware of the level of abstraction at which he operates, for there is always the danger that anyone, especially the scientist, may tend to mistake the conceptual abstraction for what it refers to; embrace it eagerly because of the feeling of security it affords; forget that it obscures uniqueness and differences, that it is a function of some human purpose, and that it is at best partial and tentative.

Any such approach to the study of man's values and ideas is automatically denied us if we follow certain current "schools" of psychology which, stemming from the scientific tradition set by Descartes, attempt to explain the nature of man entirely in terms of a mechanistic determinism. This does not in the least, of course, imply that we turn our backs on the results of rigid scientific experimentation. Indeed, the more rigorous of these as found in modern physics support the conclusion that our traditional conceptions of matter are due for serious revision; and, further and more important for us here, the conclusions of modern physics as well as modern biology deny any complete determinism and indicate that what makes up the universe, including man, is ceaseless activity, continual flow from form where the spirit of man plays a creative role.

It may be helpful for us in setting the stage if we differentiate among four different ways in which experience may be viewed, four differentiable areas of complexity.

1. Ongoing Naive Experience

This is the level of immediate, "pure" experience as experienced—unanalyzed, unconceptualized, unmediated, and with no concern on the part of the experiencing individual to describe, analyze, conceptualize, or communicate his experience. This ongoing, naive experience is what Alfred Korzybski called "first-order" or "objective level" or "un-speakable" experience. He wrote: "Thus we *handle* what we call a pencil. Whatever we *handle* is un-speakable; yet we *say* 'this *is* a pencil,' which statement is unconditionally false to facts, because the object appears as an absolute individual and *is not* words."[1] This is perhaps what William Wordsworth had in mind when he said, "We murder to dissect."

As has been frequently pointed out, any attempt to describe or analyze experience immediately alters that experience. When we are trying to describe or analyze experience or any aspect of it, we are functionally organized quite differently than when we are participating in a process of living and are not describing or analyzing it. Our experiences in the occasions of living are dependent upon and characterized by processes involving, for example, overtones of satisfaction or dissatisfaction, a sense of involvement and responsibility, a sense of intent or aim, commitment through activity, a sense of worry, frustration, or urgency, a sense of despair, hope, or faith, and so forth, depending upon the particular orchestration going on in a particular unique occasion of living. The obvious complexity of the orchestration that any ongoing experience *is* makes the process of isolating aspects of it particularly difficult. But if the psychologist ignores any significant aspects of experience in an attempt to isolate easily manipulable variables, he is bound to fail in his attempt to understand that experience.

2. Description

Verbalization and communication, either retrospective or simultaneous with the occurrence of some experience, may be distinguished methodologically from naive experience itself because of a form of awareness of selected aspects of experience. Some focusing, categorization, and coding are operating in the process of dealing consciously or verbally with any selected experiential aggregate. It is as if, with any such focusing, awareness is shifted from the full orchestration as a whole to the role of a particular instrument in the orchestra.

It is important for psychologists using descriptive material of any kind never to lose sight of the fact that reports of experience are not to be equated with experience itself. Such "protocol" data, however, do provide the psychologist with valuable raw material. This is not limited to introspective reports obtained in the laboratory or to clinical material. Some of the most penetrating descriptions of experience have been given us by poets, novelists, composers, and religious prophets. From this point of view, the psychologist concerned with understanding the full range of human experience can enormously benefit by sensitizing himself to the insights found in the "humanities." The humanists, on the other hand, should be able to profit by keeping in touch with the psychologist who, in his self-appointed capacity as scientist, must try to systematize the intuitive portrayals of humanists and formulate constructs of general validity.

3. Focused Analysis and Conceptualization

Instead of focusing on a selected phenomenon in an experience of living we may, in the midst of that living, try to "figure out" conceptually what is going on. We do so for some purpose. Perhaps we are trying to resolve some personal problem, perhaps we are delving into our own experience in the hope of discovering hunches or clues that will provide us with some hypothesis, or perhaps we are only obeying the instructions of some experimenter in the psychological laboratory. Analysis of any occurrence for whatever purpose is a very different experiential aggregate than purposive behavior itself or focused aspects of it.

This area of complexity described as "focused analysis and conceptualization" also includes our attempt to understand the behavior and purposes of other people as we try to carry out our own purposes in social situations. Such understanding will usually be successful insofar as we are able to bring to an occasion appropriate sets of abstractions derived from our own experience. Our ability to "put ourselves into another's position" and to "share" his experience vicariously seems generally to depend upon the similarity of our experiential backgrounds, our purposes, our standards for sensing satisfaction, our values, and so forth. In other instances, where there may be little similarity of experiential background, our understanding of others may increase if the situation is such that we develop with those others some community of interest through the repeated sharing of new experiences.

No matter how close the correspondence may be between all the factors involved in giving us an awareness of another person's experience under certain circumstances, we still have to interpret his experiences in terms of our own experience. And no matter how closely knit the person may feel to the social group, no matter how great the correspondence may be between his experiential background, his needs, his purposes or values and those of other members of the group, his experience of participation with the group is still uniquely his own. Complete understanding of another person is an unobtainable ideal.

Yet our understanding of another person (or his understanding of us) may be more accurate than his own understanding of himself, since our "explanation" of his behavior may be taking more factors into account, may give very different priorities to different aspects that constitute the aggregate experience, may differentiate or abstract out of his process of living as we observe it more intrinsically reasonable aspects so that we are more aware of the processes playing a role in his behavior than he is himself. Thus, an experienced physician will be able to tell us why we are tired or suffer certain aches or pains; an experienced psychiatrist may be able to point out to us what purposes of which we are unaware seem to be guiding our actions; a wise friend may be able to point out contradictions in our value standards, and so forth.

4. Abstracting for Scientific Specification

The scientist's attempt to understand the nature of human living is ultimately an attempt to distinguish components, to choose those by means of which he will be able to define and interpret the significance of any process of living, and to describe the variables on which the singularity of any process depends. If a scientist's abstracting can be effectively related to his presuppositions, then he will have an instrument to render communication more accurate and to enable others to understand the abstractions without reference to any particular item of behavior that might illustrate it.

Such scientific abstractions are not affected by individual behavior and are not altered when conceptualized from the point of view of different persons. If they were so affected or altered, they would prove useless; it is their static quality that gives them the utility they have in understanding the significance of concrete behavioral situations. This does not mean, of course, that such scientific abstractions never change. They are, on the contrary, constantly evolving and being modified by scientists themselves to increase their usefulness. The creative scientist tests his abstractions by their performance, not by their consistency, realizing that any abstraction is highly tentative. In describing William James, Alfred North Whitehead wrote, "His intellectual life was one protest against the dismissal of experience in the interest of system."[2] What we mean by the "static quality" of an abstraction is simply that scientific abstractions would be operationally useless if their significances were not "fixed."

But it should be emphasized again that when we are dealing with this fourth area of complexity which makes scientific communication possible, we are necessarily violating phenomenal data. The psychologist's awareness of this fact and of some of the omissions involved in operating in this area should help give him perspective to increase the usefulness of his abstractions.

Two points should be noted in passing with respect to this fourfold differentiation: First, the way a process of living is considered will depend upon one's purposes, and second, no matter which way a process of living is considered, the consideration is still yours, is still a process in which you cannot be detached, is still a transaction involving what Percy Bridgman has called the "personal equation" from which no scientist can escape. All experiences are from a personal behavioral center.

The first-order data of human living, whether sought by the poet or the scientist—the nature of naive ongoing experience and behavior—seem unattainable or at least unreportable. As Henri Amiel put it: "To speak is to disperse and scatter. Words isolate and localize life in a single point; they touch only the circumference of being; they analyze, they treat one thing at a time." If it is true that experiencing *as* experiencing, living *as* living, is unreportable and unanalyzable without destroying its "reality," it is perhaps true that humanism gives us the closest approach to the raw, descriptive data we desire.

A most important point, then, to understand clearly at the outset is that experience *as* experience, as it occurs independently of any purpose to describe or analyze it, is almost inaccessible both to the humanist and to the psychologist. Naive, ongoing human experience seems unreportable, ineffable. But it is the men and women we call "humanists" who seem to come closest to capturing and conveying these ineffable, unconceptualizable experiences. Hence we return to them over and over again in trying to get a toehold on our own awareness, or in trying to lift ourselves up to more satisfying levels of experience. This point is nicely illustrated by a passage from Robert Henri's *The Art Spirit:*

> There are moments in our lives, there are moments in a day, when we seem to see beyond the usual—become clairvoyant. We reach then into reality. Such are the moments of our greatest happiness. Such are the moments of our greatest wisdom.
>
> It is in the nature of all people to have these experiences; but in our time and under the conditions of our lives it is only a rare few who are able to continue in the experience and find expression for it.
>
> At such times there is a song going on within us, a song to which we listen. It fills us with surprise. We marvel at it. We would continue to hear it. But few are capable of holding themselves in the state of listening to their own song. Intellectuality steps in and as the song within us is of the utmost sensitiveness, it retires in the presence of the cold, material intellect. It is aristocratic and will not associate itself with the commonplace—and we fall back and become our ordinary selves. Yet we live in the memory of these songs which in moments of

intellectual inadvertence have been possible to us. They are the pinnacles of our experience and it is the desire to express these intimate sensations, this song from within, which motivates the masters of all art.[3]

The Greek historian Polybius (201–120 B.C.) observed that:

in their proverb "The starting point is half the whole," the Ancients recommended the payment of the utmost attention in any given case to the achievement of a good start; and what is commonly regarded as an exaggerated statement on their part really errs, in my opinion, by falling short of the truth. It may be asserted with confidence that the starting point is not "half the whole" but that it extends right to the end. It is quite impossible to make a good start in anything without, in anticipation, mentally embracing the completion of the project or realizing in what sphere and to what purpose and for what reason the action is projected. It is equally impossible adequately to summarize any given course of events without, in the process, referring to the starting point and showing whence and how and why that point has led up to the actual transactions of the moment. Starting points must accordingly be regarded as extending not merely to the middle but to the end, and the utmost attention ought, in consequence, to be paid to starting points by both writers and readers of Universal History.[4]

After the problem has been formulated, those relevant variables which seem most important are defined in such a way that they can be used as experimental variables. The next step is that of actual experimental design: working out methods for bringing the selected variables into relationships appropriate for the problem and selecting or devising techniques for manipulating the variables and interpreting the data. The experimental investigation is then carried through and scientific conceptualizations modified in the light of the experimental outcomes. The original problem can now be reformulated in the light of the new data and conceptualization, and the process of inquiry repeated. The entire process of scientific inquiry is a self-correcting, never-ending one.

Whitehead has noted that "no science can be more secure than the unconscious metaphysics which it tacitly presupposes."[5] It is obviously impossible to disclose that of which we are unaware. So we should try to spell out in some detail those presuppositions and postulates of which we are aware, with which we begin, and which shape our inquiry throughout. Postulations arise not only from a field of inquiry, but also as statements of conditions under which work is done or will be done within the field of inquiry. They are always open to reexamination.

I should like to quote once more from Polybius, to whom I am very partial. According to Toynbee, it was Polybius who coined the Greek word which Toynbee says is "exactly the equivalent to the word 'transaction.'" The point of view of psychology that a small group of us have tried to work out in the past two decades we labeled "transactional psychology," following the use of the term by John Dewey and Madison Bentley who, I am certain, would have been most excited to have known that the word was coined by Polybius for the same reasons they used it, just as I was when I read Polybius after talking for years about transactionalism.

Polybius writes:

> In previous periods the transactions of the habitable world took place in separate compartments, in which the projects attempted, the results attained, and the localities involved were unrelated. But from this date onward history acquires an organic character, the transactions of Italy and North Africa become involved with those of Hellas and Asia, and all the currents set toward a single goal....
>
> That mighty revolutionary, whose pawns are human lives, has never before achieved such an astonishing *tour de force* as she has staged for the benefit of our generation; yet the monographs of the historical specialists give no inkling of the whole picture, and if any reader supposes that a survey of the leading countries in isolation from one another, or rather, the contemplation of their respective local chronicles, can have given him an intuition into the scheme of the world in its general arrangement and setting, I must hasten to expose his fallacy. To my mind, the persuasion that an acquaintance with local history will give a fair perspective of the whole phenomenon is as erroneous as the notion that the contemplation of the *disjecta membra* of a once living and beautiful organism is equivalent to the direct observation of the organism itself in all the energy and beauty of life. I fancy that anyone who maintained such a position would speedily admit the ludicrous enormity of his error if the organism could be revealed to him by some magician who had reconstituted it, at a stroke, in its original perfection of form and grace of vitality. While the part may conceivably offer a hint of the whole, it cannot possibly yield an exact and certain knowledge of it; and the inference is that the specialists have a singularly small contribution to offer toward a true understanding of world history. The study of general contacts and relations and of general resemblances and differences is the only avenue to a general perspective, without which neither profit nor pleasure can be extracted from historical research.[6]

This organic approach described by Polybius has a very modern sound indeed, and it poses a most important problem for the process of scientific inquiry. I can state it by citing a conclusion reached by C. S. Smith, who wrote in an article entitled "Materials and the Development of Civilization and Science":

> It will be apparent from this far-too-hasty view that *in the past,* most knowledge of materials has been gained by the intelligent empiricist, and the role of science has been to explain and to provide better control rather than to open up new areas. It may be that in complex fields this is inevitable, for science invokes subtle abstractions which are necessarily concerned sequentially with parts rather than simultaneously with a whole system.[7]

It may be that scientific inquiry—especially in a field like social psychology, dealing with the complicated behavior of a human being in a social matrix—has, in its initial stages, some of the characteristics of the art of the practitioner or the innovator who finds a solution for a problem without always being completely aware intellectually of what variables he actually is taking into account. A philosopher of science, Michael Polanyi, has observed that "it is pathetic to watch the endless efforts—equipped with microscopy and chemistry, with mathematics and electronics—to reproduce a single violin of the kind the half-literate Stradivarius turned out as a matter of routine more than 200 years ago."[8]

And we can refer to Archimedes, who Plutarch tells us invented his many mechanical devices only because of King Hiero's desire "that he should reduce to practice some part of his admirable speculation in science, and by accommodating the theoretic truth to sensation and ordinary use, bring it more within the appreciation of the people in general." But, Plutarch continues, "Archimedes possessed so high a spirit, so profound a soul, and such treasures of scientific knowledge, that though these inventions had now obtained him the renown of more than human sagacity, he yet would not deign to leave behind him any commentary or writing on such subjects; but repudiating as sordid and ignoble the whole trade of engineering, and every sort of art that tends itself to mere use and profit, he placed his whole affection and ambition in those purer speculations where there can be no reference to the vulgar needs of life."[9]

The art of investigation is the art of scientific inquiry which—like any other form of inquiry—is undertaken to solve a problem. It is the way an investigator poses a problem to himself that determines what he will come out with. And this art of scientific inquiry is not something that can be neatly specified, articulated, or taught. It can, perhaps, best be acquired by close and continued apprenticeship with someone who has demonstrated the art, particularly if the apprentice has the opportunity to see how the more experienced man tackles a situation that is new to both of them and can get a sense or feel of how he formulates the problem and then goes at it.

While this is obvious enough, all too often scientific *inquiry* is confused with scientific *method,* which is a means or technique of pursuing scientific inquiry. If scientific inquiry is equated with scientific method, there is always the great danger, particularly when human beings are the material of investigation, that the investigator will leave out of his consideration the all-important problem of formulating his problem. And this means he may automatically and without being aware of it rule out of his formulation of the problem a host of functional variables that do not readily fit his scientific methodology.[10]

A particular danger, partly because it has become so fashionable, is to equate "scientific inquiry" only with quantitative procedures. The fetish for measurement can easily blind an investigator to the much more important aspect of scientific inquiry, namely, the problem of problemization, of making sure that important relevant variables are not left out. Fruitful scientific inquiry proceeds from discovering first *if* one abstracted variable is related to another; then *how* it is related; and, finally, *how much* it is related. If an investigator is not clear on the priorities of these steps and on the function of quantitative procedures, then his quantification, no matter how elaborate, may turn out to be only scientific pretension.

This is not to deny in the slightest the important and often indispensable roles of quantitative measurement in scientific methodology, such as the validation of a hunch or the establishment of norms. But naively following procedures that have proved dramatically successful in the natural sciences has produced a tremendous amount of rather sterile quantitative data in psychology, as well as in other branches of the human sciences.[11] Furthermore, in the natural sciences themselves, of course, measurement not related to unfettered speculation is equally barren.

Unless an investigator has become hardened or fixed in his approach, his style or manner or going about his work should acquire improved finesse and appropriateness to the situation at hand. His art of investigation should presumably improve as he learns through his own gropings what forces and influences have a bearing on the way he should formulate his problem and the methods he should utilize or devise to solve it. The methodological procedures are, incidentally, relatively easy to work out once the formulation has become clear.

As has so often been noted, the art of investigation and the process of scientific inquiry are never-ending processes for anyone who is plugged into the ceaseless changes around him. This is patently clear in psychology, where a dozen new problems emerge as soon as one problem is resolved. The true investigator thus becomes almost obsessed with his art and cannot get it out of his hair. When Pavlov was asked by his students how they could become as inventive as he was, he is said to have replied: "Get up in the morning with your problem before you. Breakfast with it. Go to the laboratory with it. Eat your lunch with it. Keep it before you after dinner. Go to bed with it in your mind. Dream about it."[12] An even more glaring example is recorded by Plutarch when, in discussing Archimedes, he reports that "the charm of his familiar and domestic Siren made him forget his food and neglect his person, to that degree that when he was occasionally carried by absolute violence to bathe or have his body anointed, he used to trace geometrical figures in the ashes of the fire, and diagrams in the oil on his body, being in a state of entire preoccupation, and, in the truest sense, divine possession with his love and delight in science."[13]

While I am not necessarily advocating any such extreme concentration to aspiring psychologists, I do feel it important to emphasize that an investigator can never contribute what he might to this or any other field if he allows himself to be continuously diverted or distracted. A particularly pernicious form of distraction is one that has become sanctioned, almost sanctified, these days: the participation in endless conferences, meetings, and consultantships, with their halo of prestige and their legitimation of mistaking exchanges of professional banter for creative inquiry. As a by-product, anyone dedicated to scientific inquiry must be prepared to be lonely a good deal of the time if he cuts loose and accepts the priorities for research that seem to him to present the greatest challenges, irrespective of the fads and fashions of his discipline and of the social satisfactions and rewards that come from association with his peers. After spending several hours with Einstein showing him our perception demonstrations and telling him of the resistance that I had encountered from stimulus-bound psychologists, he smiled broadly and remarked: "I learned many years ago never to waste time trying to convince my colleagues."

The major advances in science are due much more to improved formulations and the discovery of new variables than they are to improved or new methodologies, although the latter must, of course, never be underrated. The scientific inquiry that we call psychology will increase our understanding of experience and behavior insofar as it can differentiate the processes that play a role, demonstrate how the variations of certain processes affect other processes, and how all of them together transact to constitute the orchestration of biological, personal, and social living.

NOTES

1. Alfred Korzybski, *Science and Sanity: An Introduction to Non-Aristotelian Systems and General Semantics,* 3d ed. (Lakeville, Conn.: Institute of General Semantics, 1948), p. 35.
2. A. N. Whitehead, *Modes of Thought* (New York: Macmillan, 1938), p. 4.
3. Robert Henri, *The Art Spirit,* comp. Margery Ryerson (Philadelphia: J. B. Lippincott, 1923).
4. Arnold J. Toynbee, trans., *Greek Historical Thought from Homer to the Age of Heraclitus* (New York: Mentor Books, 1952), p. 137.
5. A. N. Whitehead, *Adventures of Ideas* (New York: Macmillan, 1954), p. 197.
6. Toynbee, *Greek Historical Thought,* pp. 44–46.
7. *Science,* 148 (1965):916.
8. *Personal Knowledge* (University of Chicago Press, 1948), p. 196.
9. *Plutarch's Lives,* ed. J. and W. Langhorne (New York: Thomas Crowell, n.d.), 1, pp. 486, 488–89.
10. For a more extended discussion of this subject, see Hadley Cantril, Adelbert Ames, Jr., Albert H. Hastorf, and William H. Ittelson, "Psychology and Scientific Research," *Science,* 110 (1949):461–64, 491–97, and 517–22.
11. For a penetrating discussion of the overemphasis of quantification in psychology, see Henry A. Murray, "The Personality and Career of Satan," *Journal of Social Issues,* 18, no. 4 (1962):36–46.
12. J. R. Baker, *Science and the Planned State* (London: Macmillan, 1945), p. 55.
13. *Plutarch's Lives,* p. 489.

ABNER SHIMONY
Boston University

Is Observation Theory-Laden? A Problem in Naturalistic Epistemology

> Erect your schemes with as much method and skill as you please; yet if the materials be . . . spun out of your own entrails . . . the edifice will conclude at last in a cobweb. . . . As for us the Ancients, we are content with the bee to pretend to nothing of our own, beyond our wings and our voice, that is to say, our flights and our language. For the rest, whatever we have got, has been by infinite labour and search, and ranging through every corner of nature.
>
> —Jonathan Swift
> *The Battle of the Books*

1. INTRODUCTION

The orthodox logical empiricist treatment of the relation between scientific theories and observations (as exemplified in the work of Rudolf Carnap, Ernest Nagel, Carl Hempel, and R. B. Braithwaite) abstained as a matter of principle from considerations of empirical psychology. Since psychology is the least developed of the natural sciences, an appeal to it was supposed to subvert the epistemological program of establishing a firm foundation for all the sciences. Furthermore, epistemology was not considered to be in need of the answers to the questions typically investigated by empirical psychology. It is also likely that the abstention of the orthodox logical empiricists from empirical psychology in their treatment of the relation between theories and observations was in large part due to their acceptance of Gottlob Frege's polemic against psychologism in logic, for they regarded the central problem concerning this relation to be one of logic.

One of the most influential criticisms of orthodox logical empiricism has been N. R. Hanson's argument that perception is intertwined with concepts, so that observation is usually, or in crucial circumstances, "theory-laden." Hanson cites a large number of experiments in empirical psychology in developing his argument, and these citations appear to be indispensable to his case.[1] The primary purpose of this essay is to evaluate critically Hanson's argument.[2] It is important to do so partly because of widespread acceptance of his views or similar ones,[3] but even more because of their bearing upon the general program of naturalistic epistemology. In this program, scientific results concerning the place of human beings in nature have important implications for epistemological problems. Accordingly, the abstention of the orthodox logical empiricists from considerations of empirical psychology deprived them of valuable information for epistemological investigations. However, even though I strongly concur with Hanson's view that psychology is relevant to epistemology, I disagree with some of the citations from empirical psychology which he deploys, and even more with the mode of deployment. If the disagreement only concerned the former matter, then I would be in effect asserting that epistemological problems can be settled by an appeal to psychology. Rather, I believe that there should be a dialectic interplay between psychology and epistemology, and a secondary purpose of this essay is to illustrate this interplay, even if it is difficult to articulate the process.

As regards the thesis that observation is theory-laden, I shall maintain that it is a great oversimplification of the functioning of our cognitive faculties. It seems to me that what emerges from the psychological literature on perception is a complex picture of various kinds and degrees of integration between percepts and concepts together with mechanisms for switching from close integration to relative autonomy. There are strategies of perception, which can appropriately be called "integrative," in which sensory clues seem to be suppressed from consciousness, and the resulting perceptual judgments extrapolate far beyond any input that can reasonably be characterized as "given." The existence and importance of integrative strategies give some support to the thesis that observation is theory-laden. However, there are also strategies which can appropriately be called "analytic," in which beliefs and theories are to some extent held in abeyance and clues are brought to the surface of consciousness. The neglect of these strategies is Hanson's fundamental error. I shall maintain that the ability of human beings to switch among strategies of different kinds is essential to the reliability of our cognitive processes in general, and particularly of the controlled processes of scientific investigation.

The logical empiricist thesis that there is a sharp distinction between theory and observation has been challenged from other points of view than Hanson's. Thus, P. Achinstein and F. Dretske point out that the techniques of observation, and hence the extension of the term "observation," are changed by the development of scientific theories. H. Putnam argues that there is no clear dichotomy between observational and theoretical terms, such that observation reports employ only the former.[4] These challenges are important for correcting an artificially schematized picture of scientific knowledge presented by the logical empiricists. Nevertheless, there is a core of good sense in the assumption of the logical empiricists that "publicly observable" things, properties, and events can be acknowledged by the advocates of conflicting theories. Psychological theory is able to supply a vindication (with some reservations) of their assumption, thereby supporting to some extent a position developed in deliberate abstention from psychological considerations.[5]

2. HANSON'S ARGUMENTS

In this section I shall summarize Hanson's arguments that observations are theory-laden, with particular attention to his reliance upon the results of empirical psychology. The discussion will be almost entirely concerned with visual observation, since nearly all of Hanson's examples are visual, though occasional remarks suggest that he takes essentially the same position with regard to all the senses.[6]

Although Hanson acknowledges the existence of genuine cases of phenomenal seeing, in which there is little or no conceptual element, he considers them to be atypical.[7] The typical case of seeing is *seeing that* . . . , where the ellipsis indicates a sentential clause, a clause that could stand separately as a complete declarative sentence.[8] *Seeing that* takes place even when the report of the seeing has the grammatical form "*N* sees *X*," where *X* is a common noun rather than a sentential clause, for Hanson makes the following exegesis of such reports: "What is it to see boxes, staircases, birds, antelopes, bears, goblets, X-ray tubes? . . . It is to see that, were certain things done to objects before our eyes, other things would result." *Seeing that* is the sense of seeing which Hanson considers to be crucial in research in the natural sciences, for

> 'Seeing that' threads knowledge into our seeing; it saves us from re-identifying everything that meets our eye; it allows physicists to observe new data as physicists, and not as cameras. . . . Observation in physics is not an encounter with unfamiliar and unconnected flashes, sounds and bumps, but rather a calculated meeting with these as flashes, sounds and bumps of a particular kind.[9]

Finally, it appears from Hanson's examples that the implicit sentential clauses in *seeing that* can be supplied by any part of the subject's system of knowledge and beliefs. Thus, "Tycho and Simplicius see that the universe is geocentric; Kepler and Galileo see that it is heliocentric."[10]

The argument for these theses commences with a consideration of reversible perspective drawings and ambiguous figures, which are familiar from textbooks of Gestalt psychology. In seeing one of these figures in two different ways—as bird or antelope, as convex or as concave—he says that it is natural to say that "we see different things." But the difference is neither in the way the retina is affected nor in the sensations, since the same lines are involved in the two instances of seeing.[11] The difference cannot be due to differing interpretations, because, as Ludwig Wittgenstein says, if seeing an ambiguous figure as a box means "I am having a particular visual experience which I always have when I interpret the figure as a box, or when I look at a box," then "if I meant this, I ought to know it. I ought to be able to refer to the experience directly and not only indirectly."[12] A second reason given by Hanson does not occur in Wittgenstein's text, namely, that in ordinary usage when we use the term "interpretation" we refer to a process which takes time, and this is not the case in seeing an ambiguous figure in one way rather than another; but instantaneous interpretation is inadmissible, for it "hails from the Limbo that produced unsensed sensibilia, unconscious inference, incorrigible statements, negative facts and Objektive." What is different in seeing the ambiguous figure as a bird or as antelope is the *organization,* which "is not itself seen as are the lines and colours of a drawing. . . . Yet it gives the lines and shapes a pattern. Were this lacking we would be left with nothing but an unintelligible configuration of lines." Immediately after this answer, Hanson asks a question, which sounds like a request for a causal explanation: "How do visual experiences become organized?" He answers that "the context gives us the clue" and then explains that "such a context, however, need not be set out explicitly. Often it is 'built into' thinking, imagining and picturing. We are set to appreciate the visual aspect of things in certain ways."[13] Hanson then makes reference to a large body of literature in experimental psychology on "set" and "Aufgabe."[14] He comments on this reference that "philosophy has no concern with fact, only with conceptual matters (cf. Wittgenstein, Tractatus, 4.111); but discussions of perception could not but be improved by the reading of these twenty papers";[15] and "one can talk philosophy by way of a factual discipline without thereby allowing his case to stand or fall on the success of that discipline."[16]

In both of his general books on philosophy of science, Hanson shifts to a second stage of argument in which the reliance upon evidence from empirical psychology is minimized, and at the same time his conclusions are strengthened in the sense that "an alternative account would be not merely false, but absurd."[17]

The argument proceeds as follows: (1) Scientific knowledge has an inescapable linguistic element, for it is expressible in sentences and presumably only in this way.[18] (2) A sentence is not a picture of a state of affairs, for assertion has a character—as Wittgenstein emphasized in his later work, when he rejected the views of the *Tractatus*—which is distinct from representation.[19] (3) The sense of "seeing" which is important for natural science is determined by the requirement that it be significant and relevant for scientific knowledge. But,

> if seeing were a purely visual phenomenon untainted by any of the effects of language, then nothing that we saw with our eyes would ever be relevant to what we know about the world, and nothing that we know about the world could even have significance with respect to what we say we see.... Our vision would be without understanding, our understanding without light. If this is absurd it will show again how different is seeing from the mere formation of retinal and mental pictures, and how central to the connection between vision and knowledge is the concept of *seeing that.*[20]

Hanson concludes that "it is a matter of logic, not merely a matter of fact, that seeing as and seeing that are indispensable to what is called in science, *seeing* or *observing.*"[21] Since for Hanson, who clearly follows Wittgenstein on this matter, logic is concerned with the rules of language,[22] it is important to identify the language to which he is implicitly referring when he draws his conclusion. Although his texts are not entirely clear on this point, one can reasonably say that the language he has in mind is that of ordinary discourse, with perhaps some enrichment (of which ordinary language is susceptible) with technical terminology. The evidence for this interpretation is his insistence that the second stage of his argument does not rely upon psychology or neurophysiology,[23] which surely implies that the language to which he is referring is not any theoretical language of one of these disciplines. Furthermore, he asserts that what "ordinary people mean ... is of maximum importance when we are concerned to understand a general philosophical concept like seeing."[24] The component of ordinary-language philosophy in Hanson's thought must be kept in mind even in an examination of the first stage of his argument, which explicitly draws upon some results of empirical psychology.[25]

3. THE CONFLATION OF SENSES OF "SEEING"

Hanson recognizes (with evident pleasure) the extraordinary variety of syntactical and semantic constructions of the verb "to see."[26] Two of these senses he separates from the others: seeing in the intransitive sense "indeterminately or in general, as do infants and lunatics";[27] and phenomenal seeing, which occurs in artificial situations such as an oculist's tests, or when one is confused by unfamiliar phenomena. His dominant tendency, however, is to conflate rather than to discriminate senses of the verb. Although Hanson repeatedly cites Wittgenstein on matters of philosophical method and on specific questions of analyzing locutions of seeing, he does not heed the warning of *Philosophical Investigations* (p. 66) against seeking for one essential use underlying the varied uses of a word. Indeed, at one point, Hanson explicitly takes exception to the warning. In typically exploratory and tentative manner Wittgenstein says, "Seeing as . . . is not part of perception. And for that reason it is like seeing and again not like" (p. 197). Hanson comments:

> Something of the concept of seeing can be discerned from tracing uses of 'seeing as . . .'. Wittgenstein is reluctant to concede this, but his reasons are not clear to me. On the contrary, the logic of 'seeing as' seems to illuminate the general perceptual case.[28]

In some cases the conflation of senses of "to see" is crucial to Hanson's epistemological position, but in order to avoid becoming enmeshed in peripheral issues it will be helpful to point out one case of conflation which is both unconvincing and inessential to his central thesis. In a passage quoted at the beginning of section 2, Hanson construes "*N* sees *X*"—where *X* denotes an object—as implicitly meaning "*N* sees that . . ." with the blank filled by an appropriate sentential clause. However, great confusion often results unless the context following the verb in "*N* sees *X*" is extensional. For example:

> The boy points to an X-ray tube and asks, what's that? The father answers 'that is an X-ray tube'. The answer is not only true but relevant. Yet how could it be relevant if the son did not already see an X-ray tube?[29]

For similar reasons, a language which lacked the extensional usage of "to see" would require circumlocutions in order to express the fact that two observers discriminate visually a common object which they differently describe or identify or theorize about. Except for the diminution of the rhetoric in certain passages, Hanson's exposition of his main thesis would not suffer from the recognition of an extensional sense of "to see."

The circumstances in which Hanson's epistemology requires the conflation of "seeing *X*," "seeing *X* as . . . ," and "seeing that . . ." are rather special, but in them the conflation has a *prima facie* plausibility. They are circumstances in which the locutions are either used in the first-person singular,[30] or else in the second or third person but with a sympathetic modality, as if to represent that person's point of view. First-person singular usage excludes the extensional sense of "to see" discussed in the preceding paragraph, for my naming the object will involve recognizing it, even erroneously, in terms of some identifying characteristic. Hence the thing I see is seen *as* an object having these characteristics, and I see *that* it has these characteristics and whatever characteristics I believe to be associated with them (or at least I can say this if I disregard the possibility of error). Exactly the same justification for the conflation of senses applies to second- or third-person usage with a sympathetic modality, for then there is not only an assertion of a visual discrimination but tacit reference to the mental state associated with that discrimination. Since Hanson is evidently committed to writing the history of science from the points of view of the historic figures, it is understandable that his third-person locutions should be in a sympathetic modality (and, conversely, it is understandable why his conflation of senses of "to see" has appealed to some historians of science).

A historical example, however, will undermine the plausibility of the foregoing defense of conflation. Because Hanson maintains that observation is theory-laden (the conflation of "seeing *X*" and "seeing that . . ."), he declares that "Tycho and Simplicius see a mobile sun, Kepler and Galileo a static sun."[31] If so, what did Newton see when he looked at the sun? Now Newton tells us what his dynamical theory implies about the motion of the sun: "The sun is agitated by a perpetual motion, but never recedes far from the common center of gravity of all the planets" (Bk. III, Prop. XII). But it would surely be ironical to attribute to Newton the opinion that he *saw* a sun agitated by perpetual motion, for Newton insisted upon a strong principle of visual relativity, that only relative motion among bodies could be discerned by sight; the criterion for absolute motion, which is what his theory is ascribing to the sun, is a dynamical one, applicable to phenomena only by means of the systematic combination of observation, theory, and analysis presented in the *Principia.*[32] To generalize, if a scientist's theoretical view of the world is sufficiently comprehensive to include a theory of vision, and if that theory sharply distinguishes visual evidence from cognition, then Hanson's view that seeing is theory-laden will generate a paradox, and perhaps an outright contradiction. The example suffices to show that something is amiss in Hanson's claim, but further analysis is required to diagnose the difficulty more precisely.

4. HANSON'S "LOGICAL" ARGUMENT AND ITS RELATION TO KANT

I shall first examine the later stage of Hanson's argument, which allegedly dispenses with empirical evidence and establishes a necessary connection between seeing and knowing. The argument is so loose in texture that it is hard to reconstruct it in a compelling manner. There is certainly something correct in the insistence that an integration of perception with thought occurs in our experience, at least in verbalized experience. Otherwise, he says, in a phrase that deliberately echoes Kant, "our vision would be without understanding, our understanding without light."[33] The trouble is that this dictum is a commonplace, meaning different things to different philosophers—a possibility of which Hanson should be particularly aware, in view of his general doctrine of relativity of meanings to conceptual framework. When Kant says in the "Transcendental Logic" that "intuitions without concepts are blind," he states which are the concepts that are given *a priori* and are indispensable for nonblind experience, and he provides an explicit method (strained though it may seem to us) for exhibiting these concepts systematically. Furthermore, he complicates the account by admitting judgments of perception which, in contrast to judgments of experience, do not involve the subsumption of intuitions under the *a priori* concepts, and nevertheless are apparently conscious. Hanson, of course, does not subscribe to the details of Kant's analytic, but the point is that, in the absence of a comparably detailed treatment, one cannot conclude from his argument whether there are any differentiations among the elements of a conceptual system with regard to their integration with seeing. But information on this question is indispensable to Hanson's program. If only the concepts in a universal core of common sense, or in P. F. Strawson's universal "descriptive metaphysics," or in the innate equipment which enables any normal infant to acquire a language, are integrated in a necessary manner with seeing, then the thesis of the theory-ladenness of observation loses the strong sense which Hanson evidently intends. Not only is the range of concepts integrated with seeing undetermined by Hanson's argument, but the mode of integration is entirely obscure. How tight is the integration; how much under control; how much affected by physical, emotional, and social variables? One example will suffice to show how Hanson slips from the commonplace thesis of integration to a logically unwarranted specification of it. He illustrates "bridging" by saying that in seeing objects there is "seeing that if *x* were done to them *y* would follow."[34] But visual reports can be expressed in sentential form—for example, "That appears cowlike"—without any commitments regarding the dispositions of visual objects. It may, of course, be argued that discourse in terms of appearances is derivative from discourse concerning things which have dispositions, but such an argument introduces considerations beyond the meager ones upon which Hanson has chosen to rest his case.

The foregoing remarks suggest that Hanson's position can be best defended by disregarding his claims concerning the force of the second stage of his argument, and by taking his deployment of psychological evidence as indispensable to his argument as a whole. Throughout the second stage of his argument, I suggest, he tacitly relied upon the evidence concerning Gestalt switches, illusions, and "sets" to justify his strong conclusions concerning the scope and tightness of the integration of a conceptual framework with seeing. It is necessary, therefore, to appraise his use of the evidence of empirical psychology.

It would be unwise, however, to attempt to examine the relevance of empirical psychology for all the issues Hanson has raised concerning the relation between perception and thought. Two of the issues are raised in any philosophy which is Kantian in inspiration: whether experience is possible without some minimal conceptual ordering, and whether this conceptual ordering is in some sense *a priori*. Unfortunately, the implications of empirical psychology for these issues is obscured by deep disagreements among psychologists—for instance among J. J. Gibson, the Gestalt psychologists, and Egon Brunswik on the conceptual elements of experience, and among the behaviorists, Jean Piaget, and Noam Chomsky on the question of innateness. The generically Kantian issues are so profound that they will not be illuminated by the results of empirical psychology unless the latter are organized into precise and comprehensive theories. By contrast, the special issues which Hanson raises over and above the generic Kantian ones are less profound. There are interesting psychological findings, acknowledged by adherents of different schools, that bear upon Hanson's epistemological theses. As far as possible, I shall rely upon these results in appraising Hanson's theses, though occasionally (especially on the question of unconscious inference) controversial matters will not be avoidable.

5. PSYCHOLOGICAL CONSIDERATIONS: GESTALT SWITCHES AND ILLUSIONS

This section will be devoted to collecting and commenting on some findings about ambiguous figures and perceptual illusions that are relevant to Hanson's thesis that observation is theory-laden.

Hanson contends that the characteristics of the perception of ambiguous figures reveal something important about scientific observation ("the two astronomers are to the sun as you and I might be to the duck-rabbit when you see only a duck and I only a rabbit").[35] The unnaturalness of the ambiguous figure is compensated for by the fact that it provides a kind of laboratory control: namely, one can be sure that the images on the retinas of the two observers are virtually identical, and hence that their different identifications cannot be attributed to differences in physical details, but rather to differences in organization. It is important, then, to determine empirically what factors influence perceptual organization of an ambiguous figure. One of the most striking of the psychological findings is the resistance of the Gestalt perception to conceptual suggestion. When R. Leeper's famous wife–mother-in-law figure is seen as the former by the subject, the verbal communication to him by the experimenter that the latter is also to be seen does not help him make a Gestalt switch, even though there is no question of doubting the experimenter's word. Verbal instructions about where in the drawing to look in order to pick out features of the mother-in-law's physiognomy help more than the simple conceptual suggestion; and much more helpful yet is a graduated series beginning with a drawing in which the features of the mother-in-law are unambiguous and concluding in the ambiguous drawing under observation.[36]

An even more striking example of the subject's resistance to seeing as his conceptual framework would suggest is the variant of Leeper's experiment by W. Epstein and I. Rock. The subject is shown an alternating sequence W-M-W-M-W-M-W-M of drawings which are unambiguously "wife" and "mother-in-law" respectively, and a "wife" is anticipated at the next turn; but when the ambiguous drawing is then shown, it is seen as the mother-in-law: "a clear victory for recency over expectation," and one could add that it is also a clear victory of a nonconceptual component in Gestalt organization over a conceptual component.[37]

Hanson analyzes one perceptual illusion in detail, namely the distorted room of A. Ames, in which objects in the room appear to be abnormally large or small against a background of apparently normal rectangular walls, whereas in fact the objects are of standard size and the walls are trapezoidal and tilted. He cites with apparent approbation Ames's explanation "that we wish to keep our environment relatively stable if possible, so that we can move about in it with confidence and surety" and "that we are always construing the visual stimulus pattern in terms of past experience and knowledge as well as present and future needs."[38] One further empirical fact about the experiment of Ames, however, throws great doubt on Hanson's interpretation of it: The illusion persists even after the subject has examined the construction of the room and is fully aware of the trick that is being played upon his visual system, and in fact even when the subject is a sophisticated psychologist with a theory about the mechanism of the illusion.[39] Ames's experiment allows the possibility that some elements of knowledge influence perceptual illusions, but shows that some, which *a priori* might be expected to be relevant, have no influence.

A body of data that can reasonably be interpreted to support Hanson's position concerns illusions of length and angle in plane figures, such as the arrow illusion of Müller-Lyer, the railway lines illusion of Ponzo, and the fan illusion of Hering.[40] The occurrence of these illusions can be understood as a by-product of the propensity of the apparent linear dimensions of an object to increase roughly in proportion to its apparent distance, if the retinal image is kept unchanged in size as the distance cues vary. This propensity, known as Emmert's law,[41] is evidently important for maintaining the apparent constancy of size of objects which are in motion relative to the subject. In order to apply this law to the illusions in question, it must be supposed that the figures are viewed as perspective drawings, so that, for example, converging straight lines are viewed as parallels receding into the distance.[42] The closeness of corresponding points on the converging lines then serves as a distance cue for an object lying between those points, and Emmert's law then implies that the apparent size of the object increases with closeness of the two points.

If the foregoing interpretation is accepted, then the geometrical illusions support Hanson's contention that there is "a propensity to 'alter' the details of one's visual field in order to get things sorted out intellectually,"[43] and it does so whether the achievement of constancy scaling is considered to be learned, as the authors cited maintain, or partially innate. Even more favorable to Hanson's thesis that observation is theory-laden is one further finding: that the susceptibility to the geometrical illusions decreases with age, beginning at age four or age six, according to two different studies,[44] which permits the interpretation that the acquisition of knowledge regarding spatial relations affects visual perception. If this interpretation is correct, then (in contrast to the situation of Ames's experiment) there is integration of knowledge with perception at two levels—the rather primitive level at which constancy scaling is acquired, and a more sophisticated level at which compensation is made for some of the errors caused by constancy scaling.

A quite different interpretation, however, can be given to the diminution of the geometrical illusions with age: namely, that maturation brings greater power to analyze a visual scene in a systematic manner, so that the visual impression is not controlled by what Piaget calls "centration."[45] Piaget himself suggests that decentration is intellectual rather than perceptual in character, even though it has consequences in controlling the subjectivity of perception. I do not wish to rely upon Piaget's complex and controversial theory at this point, and only mention it in order to indicate the existence of an interpretation of the empirical data quite different from that which supports Hanson.

For the purpose of this essay it suffices to consider another visual illusion which, like the geometrical illusions, is probably the by-product of constancy scaling, but which clearly resists relevant sophisticated knowledge. This is the illusion that the moon is larger when it stands near the horizon than when it is higher in the sky. Distance cues, such as occlusion and shift of angle consequent upon movements of the subject, are operative relative to terrestrial landmarks only when the moon is near the horizon,[46] and these cues combined with Emmert's law yield the illusion. Unlike the case of the geometrical illusions, however, the variation in apparent size of the moon does not seem to be affected by maturation or by theoretical knowledge. The moral is that discrimination and qualifications are necessary for a correct theory of the integration of a conceptual framework with perception. It is worthwhile to quote Rock at some length for his articulation of this moral, whether the details of his theory of perception are correct or not:

> Knowledge as such does not seem to affect perception. Yet it must be acknowledged that perception seems to move in the direction of veridicality, reflecting what has been learned about the environment. This is a problem that should not be glossed over. Still there is a difference between perceptual change produced by certain specific experiences and perceptual change produced by mere knowledge. . . . The core of the answer I am suggesting then is that when a concrete

trace containing *visual* information accrues to a specific stimulus, it can affect the way it appears. This is not the same as "knowledge about the situation."[47]

6. PSYCHOLOGICAL CONSIDERATIONS: INTEGRATIVE AND ANALYTIC STRATEGIES OF PERCEPTION

The empirical data which are most significant for evaluating Hanson's epistemological thesis concern the perception of ordinary objects, and for the most part they are not very esoteric data. An evident characteristic of the stimuli which the environment presents to our senses under ordinary circumstances is their immense richness, both in actual presentation and in potentiality for further presentations to the active observer. The stimuli presented by ambiguous drawings and typical situations of visual illusions are very meager, whereas, by contrast, the abundance of detail in ordinary circumstances makes it imperative for the organism to filter the stimulus information and make discriminations. One source of richness is the simultaneous involvement of several senses. Another is the array of "higher order variables of stimulus,"[48] such as spatial and temporal gradients, which are capable of conveying decisive information. Finally, in ordinary situations there are usually opportunities for exploration, by motion of the organism as a whole or by movements of the eyes, hands, and head, for the purpose of bringing small clues into prominence and achieving new perspectives. The actual and potential richness of the field of sensory stimuli provides an opportunity for a diversity of strategies of perception and helps to explain why it is possible, and biologically desirable, that perception be a wonderfully flexible activity.

Two major classes of perceptual strategies can be recognized, without precluding a finer taxonomy and intermediate types: one which may be called "integrative" and the other "analytic." This dichotomy seems to be recognized, with various names and descriptions, by psychologists holding very different theories. For example, for Bruner, the integrative strategy involves a complex decision process in which clues are sought, a tentative categorization is made and then checked by examining other clues, and finally an identification is performed which terminates search—and the details of this process are normally subconscious. The analytic strategy, by contrast, is a "constant close look," with conscious attention to the clues and their compatibility with available categories, and with the possible option of introducing new categories.[49] J. Gibson distinguishes between "perception," which is the gathering of information via the senses about objects and events in the world, and "sensation," which is not a preliminary process to such information-gathering, but is rather the direction of the subject's attention upon his own sensory reactions.[50] Without minimizing the deep theoretical differences between Gibson and Bruner, one can say that they agree in ascribing functional and biological primacy[51] to the integrative strategies; and I think that it is fair to make the same ascription to most other psychologists, even to the classical empiricists who maintain that sensations are constitutive of and temporally antecedent to integrated perceptions.

The thesis of the theory-ladenness of perception is exemplified very well in certain types of integrative perception, but it does not do justice to the detailed taxonomy of the entire class. Consider first the perceptual identification by means of visual clues of objects which have been experienced by various senses, for instance, different kinds of fruit. The perceptual judgment that an object is an orange is normally immediate (not preceded by deliberate conscious scrutiny of the color and visual texture prior to the judgment), and it is integrated with an expectation of taste which could never be extracted from the visual clues alone. By stretching the term "theory" to apply to knowledge of any general proposition, even of a low level of sophistication, one has here a good example of theory-laden perception. Now consider, by contrast, the perceptual discrimination of male from female chicks, whose genital eminences look very similar to a novice.[52] To make the discrimination accurately (as high as 99.5 percent) learning is essential, and the perceptual judgment of the expert is integrative in the sense that clues are not antecedently consciously discriminated. There is, however, a difference between this case and the preceding one, which is at least as important as the common denominator of requiring a background of learning. In the latter case the information for distinguishing between male and female chicks (though, of course, not all the biological potentialities associated with the distinction) is already present in the total visual stimulus, but it is there subtly as a "higher-order variable," namely a pattern. The existence of such higher-order variables is an instance of the previously noted characteristic of perception under ordinary circumstances, that the stimulus is immensely rich in details. Perceptual learning in this case consists of acquiring the skill to discriminate the pattern. In the case of visually perceiving the orange as a fruit with an expected flavor, the visual stimulus does not contain the flavor as a higher-order variable.[53] The difference between these two cases of integrative perception is blurred when Hanson draws the following parallel: "The microscopist sees coelenterate mesoglea, his new student sees only a gooey, formless stuff. Tycho and Simplicius see a mobile sun, Kepler and Galileo see a static sun."[54] He has conflated pattern recognition, in which higher-order variables in the visual stimulus are discerned as a result of training, with a type of cognitive act in which the elements integrated with visual clues are not themselves visual variables.[55] The terms "perception" and "observation" can be used broadly so as to apply to this type of cognitive act, and the advantage of doing so in describing the activity of scientists is great:

> Once the theory has achieved a status where it is no longer in the process of confirmation . . . then it is free to augment the range of observation. It gets swallowed up, so to speak, in proto-knowledge and new background beliefs and a new plateau of observation is achieved. . . . What we know is dependent, to a greater or less degree, on what we have seen to be the case; but what we are capable of seeing, epistemically, is expanded by the accumulation of more information. The greater our experience, the more inclined we are to take certain things for granted; the more inclined we are to take things for granted (providing always, of course, that they are true) the more pregnant with information become the things we can see to be the case.[56]

However, such breadth of usage is harmful if it is associated with a coarsening of the phenomenological description of perception, which can blur epistemologically important distinctions and hamper our understanding of perceptual mechanisms.

The consideration from empirical psychology which is most damaging to the thesis that observation is theory-laden is the existence of analytic strategies of perception. These strategies are frequently adopted when observers find discordance between sensory clues and the expectations suggested by their beliefs and theories. It is indeed fair to say that the availability of analytic strategies to an observer is an index of his capability to relax theoretical preconceptions.

One striking example will show how Hanson overlooks the role of analytic strategies. He cites the experiment of Bruner and Postman in which subjects are shown in tachistoscopic exposure a nonstandard playing card, such as a red six of spades, and often report seeing a black six of spades or a red six of hearts, thereby integratively perceiving the card in accordance with a normal category. Hanson fails to mention, however, some of the further results of the experiment. When the exposure time was increased, more than half of the subjects frequently experienced perceptual disruption, that is, "a gross failure of the subject to organize the perceptual field at a level of efficiency usually associated with a given viewing condition," and made comments like "I don't know what color it is now or whether it's a spade or a heart. I'm not even sure now what a spade looks like!" And upon even longer exposure most of the subjects (89.7 percent at one second, the longest exposure used) finally recognized the characteristics of the nonstandard card. A "close look," or adoption of an analytic strategy of perception, succeeded finally in disentangling features which are conjoined in the usual classification of playing cards, and which therefore are inseparable when this classification has been internalized and an integrative strategy of perception is followed.[57]

An answer on Hanson's behalf to the charge that he has neglected the role of analytic strategies of perception might be given by pointing to those passages in which he acknowledges the occurrence of phenomenal seeing, particularly in situations of confusion and conceptual muddle:

> In microscopy one often reports sensations in a phenomenal, lustreless way: 'it is green in this light; darkened areas mark the broad end. . . .' So too the physicist may say: 'the needle oscillates, and there is a faint streak near the neon parabola. Scintillation appear on the periphery of the cathodescope.'[58]

This answer, however, is inadequate for two reasons.

First, it misses the importance of an analytic strategy of perception when adherents of conflicting theories examine the results of an important experiment. Neither party may be in a state of confusion (at least subjectively), and therefore Hanson's account provides no reason for either to forgo an integrative strategy of perception in making his observations. There would then be the danger of the hedging of theories against adverse empirical data which I. Scheffler and C. Kordig have recognized in Hanson's epistemology.[59] What opposing theorists *should do,* in accordance with C. S. Peirce's admonition not to block the road to inquiry, and what they very often *in fact do,* because they are interested in finding out the truth about nature, is to adopt an analytic strategy of perception. Thus, bubble chamber photographs can be scanned cooperatively, with detailed attention to such isolable features as the thickness, length, and curvature of a track, and the number of tracks emerging from a vertex, in abstention from integratively perceiving the pictures as characteristic of certain elementary particle processes. To be sure, a communion of analytic scrutiny of experimental results is far from a sufficient condition for adjudication of theoretical disagreements. The questioning of auxiliary hypotheses provides immense resources for defending a favored theory against unfavorable data, as P. Duhem especially emphasized, and these resources can be expanded by exploiting statistical uncertainties. Stubbornness need not take the crude form of refusing to see what is plainly before one's eyes.

The second reason is more complex and has to do with the potentialities for switching to an analytic strategy of perception which are implicit in most instances of integrative strategy. A perceptual judgment resulting from an integrative strategy is potentially subject to being wondered about, doubted, censored, and reconstructed in memory as a consequence of subsequent experience and thought of the subject. He therefore has dispositions for a diversity of perceptual responses to any recurrence of the physical stimuli which resulted in the initial perceptual judgment, and among them are dispositions for adopting an analytic strategy, especially if the intervening experience should throw doubt upon the tacit premisses of the initial integration. For example, an intense auditory stimulus may result, because of an integrative strategy, in a perceptual judgment that a firecracker has exploded; but given appropriate experience or beliefs, the occurrence of an almost identical stimulus at a later time could result in a perceptual judgment of a grenade attack, or in the analytic strategy of hearing the noise and holding its interpretation in abeyance. The potentiality for such a switch to an analytic strategy of perception retrospectively throws light upon the initial perceptual judgment. It shows something about the elements of that judgment and the way in which they have been integrated, in other words, about its *internal structure.* (An analogy to a realistic treatment of the dispositions of physical objects is obvious. The disposition of a crystal to melt when heated is understood in terms of its internal structure, and conversely, the internal structure of the crystal can be probed by subjecting it, or replicas of it, to a variety of external conditions, so that various dispositions are actualized.) The possibilities for switching to an analytic strategy of perception differ greatly in the face of different types of integration, and some, such as the temporal integration of the succession of presentations during the specious present, may be greatly resistant to an analytic strategy. However, the instances of integration which are Hanson's central concern, namely, of sensory clues with conceptual elements from theories of considerable scientific sophistication, are evidently associated with strong potentialities for switching to an analytic strategy. The potentialities are easy to discern, just because the content of sophisticated theories, the evidence for them, their logical organization, and the procedures for learning them are quite well articulated; whereas the "theories" (if the term is proper) of everyday life are so deeply engrained by heredity or infantile experience that their articulation requires great effort.

7. INTERPRETATION AND THE INTERNAL STRUCTURE OF PERCEPTION

The analysis of the preceding section skirted Hanson's contention that the integration of conceptual elements into perception could not be ascribed to interpretation. This topic deserves a separate treatment, since it involves questions of the mechanism of perception more crucially than those discussed previously.

One may wonder why Hanson so insistently separates his epistemological position from the tradition represented by H. Helmholtz and more recently by Brunswik and Bruner, in which perceptual integration is considered to be a quasi-rational process, characterized variously as interpretation, unconscious inference, or probabilistic decision-making. There appear to be two philosophical reasons for Hanson's unwillingness to ally himself to this tradition. The first is his belief that the processes, postulated in this tradition, of unconscious inference and instantaneous interpretation (where "instantaneous" is applied broadly to any event with a temporal duration too brief to be consciously discerned) have very dubious ontological status.[60] The second reason is his suspicion that the tradition gives epistemological primacy to sense data or some other uninterpreted components of perception, thereby falsifying the actual phenomenology of experience and incurring the danger of excluding the subject from a public world.[61]

The first of these reasons would be expected from a philosopher of science like E. Mach, who thinks it worthless or harmful to attribute to natural things an internal structure which cannot be directly discerned; but it seems out of place in a philosopher who rejects a fictionalist understanding of the formalism of elementary particle theory.[62] A possible move at this point is to admit the meaningfulness of a microscopic description of physical, but not of psychological, entities, and in fact there is some textual evidence that Hanson does dichotomize in this way:

> While what ordinary people usually mean when they speak of electrons, or waves, or genes, or functions is of little or no importance to the understanding of such purely scientific concepts, it is of maximum importance when we are concerned to understand a general philosophical concept like *seeing.*[63]

This passage suggests that Hanson subscribes to Wittgenstein's rejection of the comparability of psychology and physics.[64] However, to deny an internal structure to a well-integrated perceptual process is grossly incompatible with the psychological evidence. One could as well maintain in the face of biological evidence that blood is a homogeneous substance. That perceptual processes admit of a meaningful "microscopic" description is initially made plausible by the fact that they are inseparable from neurophysiological processes of great rapidity and fantastic complexity. This *prima facie* argument is reinforced by a large body of psychological findings on such matters as scanning under various types of instructions, search and reaction times, suppression of visual sensitivity during saccadic movements, rhythmic patterns in auditory recognition, and so forth. Two quotations will indicate the complexity of the evidence as well as reasonable interpretations:

> Perception generally does seem to have the redundancy, wastefulness, and freedom from gross misrepresentation that characterize a parallel process. . . . Moreover, one might expect a parallel process to resist introspection, since so much unrelated activity is going on simultaneously.[65]

> Further, the result of each process must be stored or temporarily maintained, so that the information-processing sequence can be subdivided into stages, stores, and processes—each with its own sequences, time constants, and interactions.[66]

Finally, without such considerations of internal structure it is hard to make sense of the fact that the perceptual system is "tunable" with great sensitivity in extracting information from the enormously rich array of sensory input.

An answer to the second of the reasons for Hanson's aloofness from Helmholtz and his followers is that their position does not essentially require the attribution of epistemological primacy to uninterpreted components of perception. They could recognize, as Bruner explicitly does, that under normal conditions it is the perceptual judgment issuing from an integrative strategy which is the basis for articulated reports about the environment and the guide to action. The elements which are integrated—whether they are sensory clues or less familiar elements—cannot by themselves serve these purposes. When one examines the efforts of epistemologists like B. Russell, H. H. Price, and A. J. Ayer to find an error-free foundation for empirical knowledge in uninterpreted sense data, the object of their quest seems elusive, because propositional formulation and articulation require some kind of cognitive integration, thereby reintroducing the possibility of error. The stages and subprocesses in the internal structure of perception, which were hypothesized in the preceding paragraph, are more remote from consciousness than sense data and even less capable of guiding action and serving as the basis for articulated reports. To demand that they do so would be as unreasonable as demanding that a complete calculation be performed by each distinct electronic process in a computer.

What is essential to the tradition of Helmholtz is first that the elements integrated in a perceptual judgment can be discerned if the subject switches to an analytic strategy, and second that a (quasi-)rational process, inferential or interpretive, can in principle be reconstructed in terms of these elements so as to yield the perceptual judgment as its conclusion. (The qualification "quasi" is needed in order to acknowledge the likelihood that high critical standards of logic will not be satisfied by the untrained operations of human faculties.) The psychological evidence bearing on these two propositions is very complex and far from conclusive, as can be seen by considering a single area, pattern recognition. Those theories which stress feature analysis effectively incorporate the two essential propositions of Helmholtz's position, and indeed there is much evidence to "support the view that pattern recognition involves some kind of hierarchy of feature-analyzers."[67] On the other hand, much of the empirical data indicate that holistic and constructive operations occur which are akin to visual imagination,[68] and this thesis seems to be in conflict with those two propositions. The present question is bound up with the amenability of any instance of perceptual integration to an analytic treatment, and, as indicated in the preceding section, there are varying degrees of resistance to the adoption of an analytic strategy of perception. It was also pointed out there, however, that the integrations with which Hanson is concerned, involving concepts from sophisticated scientific theories, seem to be unusually amenable to an analytic strategy and also unusually rational in structure, and therefore for them the epistemological claims of Helmholtz and his school are quite reasonable.

The epistemological implications of the internal structure of perception extend beyond the question of unconscious interpretation, and also beyond the scope of this essay. A crude analogy may be suggestive. From a naturalistic point of view the human perceptual apparatus is an instrument for obtaining information about the environment, and knowledge of its principles of construction and operation, like comparable knowledge of any scientific instrument, permits us to know how reliable it is, to what errors it is subject, and what precautions may be taken to correct these errors.[69] As an illustration, the relation between integrative and analytic strategies of perception can be reconsidered. The information implicit in sensory stimuli is masked by various kinds of distortion and "noise," both external and internal to the organism. Integrative strategies of perception are particularly well adapted to overcoming certain defects inherent in the sense organs or distortions due to the position of the subject relative to objects of interest. For example, the temporal integration of successive presentations to the eyes, which are subject to spontaneous saccadic movements, counteracts the effects of the blind spots of the retinae and of the angular limits of the field of vision. The integration of visual clues with knowledge of the appearance of common objects is the fundamental technique for deriving information about interesting things which are distant from the subject. But the integrative mechanism is itself a source of possible distortion, especially in the second of the two examples just mentioned, in which the "unconscious inference" from a visual clue to a perceptual identification can be erroneous. Hence, an analytic strategy is valuable for examining and correcting the result of an uncritical integrative strategy. It is fair to say that the general reliability of the human perceptual system, which is truly astonishing if one considers the multiplicity of sources of "noise," is largely due to a metastrategy of switching between integrative and analytic strategies.

8. REMARKS ON LOGICAL EMPIRICISM

In conclusion I shall consider to what extent the foregoing criticism of Hanson confirms the orthodox logical empiricists' position concerning observation. In view of the complexities of perception that are exhibited in even an incomplete and oversimplified survey, their neglect of empirical psychology appears to be unjustifiable. How many difficulties are disregarded in a characterization like the following?

> The terms included in the observational vocabulary must refer to attributes (properties and relationships) that are directly and publicly observable—that is, whose presence or absence can be ascertained, under suitable conditions, by direct observation, and with good agreement among different observers.[70]

Surely empirical psychology supports H. Putnam's remark that "'Being an observable thing' is, in a sense, highly theoretical."[71] Nevertheless,

I claim that the orthodox logical empiricists are partly vindicated by the empirical psychology which they neglected.

What psychology in its sophistication recovers is that the naive assumption of a "publicly observable" world is largely correct. The recovery is in no sense a scientific resurrection of the epistemological thesis of direct realism, for it is accomplished by the information-processing theory of perception summarized in the preceding section, which is closer to critical realism than to any other classical epistemological position. Under suitable circumstances, perception is usually veridical in the pragmatic sense of concurrence with subsequent impressions and reliability as a guide to action. If a group of observers, not under emotional stress and with sense organs free from major defects, is placed in circumstances of good illumination, adequate opportunity for inspection, and so forth, then the perceptions of each can be expected to be veridical in the pragmatic sense, and public observation is achieved. Indeed, rough concurrence with the responses of other observers is implicit in the phrase "reliable as a guide to action." Under these circumstances an integrative strategy of perception, yielding an identification of an object or event, is usually successful, so that the agreement among observers can be readily articulated in a physicalistic vocabulary. However, the perceptual mechanism permits flexibility on this point. If there are discrepancies among the reports of the observers, a switch to an analytic strategy of perception is possible, and disagreements can be adjudicated by reference to discriminated features. The ensuing dialectic among the observers may also bring out into the open previously unarticulated disagreements in their premisses, and the result may be a retreat by one or more of the group to more cautious integrative strategies of perception. Flexibility of this sort in achieving public observation could probably be acknowledged by the orthodox logical empiricists without essentially changing their epistemological position. Had Carnap's suggestion that the observation language can be either physicalistic or phenomenalistic[72] been modified to indicate the virtues of combining both, the resulting recommendation would have agreed well with the above discussion of integrative and analytic strategies of perception.

This sketch of a partial vindication of the logical empiricists' position on observation is not intended as an endorsement of their epistemology as a whole. From the standpoint of a naturalistic philosophy, the attempt to understand theoretical terms exclusively by means of logical relations to observation terms is inadequate. What is needed is an analysis of the causal relations between those entities referred to in scientific theories and the discriminations by which human beings can become aware of these entities. Such an analysis evidently requires scientific knowledge of the place of human beings in the world and of the characteristics of their perceptual apparatus. But it also requires philosophical investigations which are not simply subsumable under the natural sciences: specifically, concerning the concept of causality itself, and concerning the applicability of the semantic notion of reference to terms in theories which are only approximately true. A number of philosophers of science are pursuing programs like the one envisaged here, and there have already been some excellent results.[73]

NOTES

This essay is an expansion of a lecture entitled "Theory, Observation, and Common Sense" delivered in 1969 at the Center for Philosophy of Science of the University of Pittsburgh. That lecture, in turn, was an expansion of part of a talk, "Proposals for a Naturalistic Epistemology," given at a workshop of the Center in 1965. Two other publications (see additional references) based upon the workshop talk present theses relevant to the present essay. I am deeply grateful to my friends Howard Stein and Richard Burian for their constructive criticisms of earlier drafts.

1. N. R. Hanson, *Patterns of Discovery* (Cambridge: Cambridge University Press, 1958), and *Perception and Discovery: An Introduction to Scientific Inquiry* (San Francisco: Freeman, Cooper, 1969). His claim that considerations of empirical psychology are only ancillary to a logical analysis of sentences expressing perceptual judgments will be examined in section 4.
2. Kordig's analysis ("The Theory-ladenness of Observation," *Review of Metaphysics,* 24 [1971]:448–84) is penetrating on questions of linguistic usage and scientific methodology, but it is little concerned with psychological questions. There have also been several good discussions of particular aspects of Hanson's position, for example, F. Dretske, *Seeing and Knowing* (Chicago: University of Chicago Press, 1969); I. Scheffler, *Science and Subjectivity* (Indianapolis: Bobbs-Merrill, 1967), p. 14; and R. A. Putnam, "Seeing and Observing," *Mind,* 78 (1969):493–500.
3. For example, T. Kuhn, *The Structure of Scientific Revolutions* (Chicago: University of Chicago Press, 1962); S. Toulmin, review of Hanson's *Patterns of Discovery,* in *British Journal for the Philosophy of Science,* 10 (1959–1960):346–49; G. Gale and E. Walter, "Kordig and the Theory-Ladenness of Observation," *Philosophy of Science,* 40 (1973):415–32; F. Suppe, *The Structure of Scientific Theories* (Urbana: University of Illinois Press, 1974); and M. Polanyi, *Personal Knowledge* (New York: Harper and Row, 1964), p. 101.
4. P. Achinstein, *Concepts of Science* (Baltimore: Johns Hopkins Press, 1968); Dretske, *Seeing and Knowing;* and H. Putnam, "What Theories Are Not," in *Logic, Methodology and Philosophy of Science,* ed. E. Nagel, P. Suppes, and A. Tarski (Stanford: Stanford University Press, 1962), pp. 240–51.
5. In the last section of this essay, the qualifications of this sentence will be explained and some criticisms of logical empiricism will be briefly stated.
6. For example, see his *Perception and Discovery,* pp. 73, 79.
7. *Patterns of Discovery,* p. 20.
8. Ibid., pp. 24–25.
9. Ibid., pp. 20–21, 22, 24.
10. *Perception and Discovery,* p. 102.
11. *Patterns of Discovery,* pp. 11–12.
12. Wittgenstein, *Philosophical Investigations* (Oxford: Basil Blackwell, 1953), p. 194, quoted in Hanson, *Patterns of Discovery,* p. 10.

13. *Patterns of Discovery,* pp. 10, 13, 15.

14. Ibid., note to p. 15, on pp. 180–81; for more detail, see *Perception and Discovery,* pp. 161–66.

15. *Patterns of Discovery,* p. 181.

16. *Perception and Discovery,* p. 131.

17. *Patterns of Discovery,* p. 24; *Perception and Discovery,* p. 131.

18. *Patterns of Discovery,* p. 25; *Perception and Discovery,* pp. 126–27.

19. *Patterns of Discovery,* pp. 26–29.

20. Ibid., p. 132; see also *Perception and Discovery,* p. 86.

21. *Perception and Discovery,* p. 147.

22. For example, ibid., pp. 184–85.

23. Ibid., pp. 130, 132.

24. Ibid., p. 79.

25. An important departure of Hanson from Wittgenstein is his extrapolation of the analysis of Gestalt switches to situations in which the subjects hold different sets of beliefs concerning matters of theoretical physics. Since Wittgenstein believed that theoretical science is a different "language game" from ordinary discourse, it seems reasonable to suppose that he would object to this extrapolation.

26. A fine collection of various constructions of "to see" is given by G. Warnock, "Seeing," *Proceedings of the Aristotelian Society,* 55 (1954–1955):201–18.

27. *Perception and Discovery,* p. 107.

28. *Patterns of Discovery,* p. 19.

29. R. A. Putnam, "Seeing and Observing," p. 494. See also Dretske, *Seeing and Knowing,* p. 37; and A. Collins, "The Epistemological Status of the Concept of Perception," *Philosophical Review,* 76 (1967):436–59.

30. See Warnock, "Seeing," p. 204, on the consequences of restricting attention to the first-person singular in analyzing "to see."

31. *Patterns of Discovery,* p. 17.

32. Bk. I, Scholium to the definitions; see also H. Stein, "Newtonian Space-Time," *Texas Quarterly* (1967):174–200.

33. *Patterns of Discovery,* p. 26.

34. Ibid., p. 30.

35. *Perception and Discovery,* p. 107.

36. U. Neisser, *Cognitive Psychology* (New York: Appleton-Century-Crofts, 1967), p. 61.

37. Ibid., pp. 141–42.

38. *Perception and Discovery,* pp. 157, 158.

39. See J. Bruner, "On Perceptual Readiness," *Psychological Review,* 64 (1957):123–52.

40. See, for example, R. Gregory, *Eye and Brain* (New York: McGraw-Hill, 1966), pp. 136 ff.

41. Dr. John Heffner has pointed out to me that Emmert's own statement of his law concerned only the apparent size of after-images, but I am following Gregory's wide application of the term "Emmert's law."

42. Gregory (*Eye and Brain*) summarizes much evidence in favor of this hypothesized connection between constancy scaling and the geometrical illusions. For example, he reports a case of a man blind since infancy whose sight was restored by corneal grafts and who acquired the ability to recognize by sight objects which he knew by touch, but whose visual depth perception and ability to identify two-dimensional representations of objects remained very faulty. This man exhibited little or no susceptibility to the geometrical illusions involving two-dimensional figures, though he was normally susceptible to the Ames illusion, in which the experimental arrangement is three-dimensional.

43. *Perception and Discovery,* p. 154.

44. M. Segall, D. Campbell, and M. Herskovits, *The Influence of Culture on Visual Perception* (Indianapolis: Bobbs-Merrill, 1966), pp. 196–97.

45. J. Piaget, *The Mechanisms of Perception,* trans. G. Seagrim (New York: Basic Books, 1969), pp. 70–71 and passim.

46. L. Kaufman and I. Rock, "The Moon Illusion," *Scientific American,* 207 (July 1962):120–30.

47. I. Rock, *The Nature of Perceptual Adaptation* (New York: Basic Books, 1966), pp. 264–65.

48. J. Gibson, *The Perception of the Visual World* (Cambridge, Mass.: Riverside Press, 1950), p. 3.

49. Bruner, "On Perceptual Readiness."

50. *Perception of the Visual World,* and *The Senses Considered as Perceptual Systems* (Boston: Houghton Mifflin, 1966).

51. I give an argument from an evolutionary point of view in favor of this primacy in "Perception from an Evolutionary Point of View," *Journal of Philosophy,* 68 (1971):571–83.

52. E. Gibson, *Principles of Perceptual Learning and Development* (New York: Appleton-Century-Crofts, 1969), p. 6.

53. A subtle intermediate case, which is not fully understood, involves intermodal transfer between senses. See Gibson, *Principles of Perceptual Learning,* pp. 215 ff.

54. *Patterns of Discovery,* p. 17.

55. Hanson also conflates pattern recognition in ordinary circumstances with Gestalt switches, but it is an essential feature of the latter that the artificially drawn ambiguous figures lack appropriate higher-order variables for resolving the ambiguity (see J. Gibson, *Perception of the Visual World,* p. 211).

56. Dretske, *Seeing and Knowing,* pp. 256–57.

57. J. Bruner and L. Postman, "On the Perception of Incongruity: A Paradigm," *Journal of Personality,* 18 (1949–1950):208–23.

58. *Patterns of Discovery,* p. 20.

59. Scheffler, *Science and Subjectivity;* Kordig, "The Theory-ladenness of Observation."

60. *Patterns of Discovery,* p. 10.

61. *Perception and Discovery,* pp. 70–74.

62. N. R. Hanson, *The Concept of the Positron: A Philosophical Analysis* (Cambridge: Cambridge University Press, 1963), pp. 46–47.

63. *Perception and Discovery,* p. 79.

64. Wittgenstein, *Philosophical Investigations,* p. 232.

65. Neisser, *Cognitive Psychology,* p. 74.

66. R. Haber and M. Hershenson, *The Psychology of Visual Perception* (New York: Holt, Rinehart, and Winston, 1973), p. 175.

67. Neisser, *Cognitive Psychology,* p. 85.

68. Ibid., chap. 4, esp. p. 95.

69. By elaborating this remark, I believe, one can refute the contention of D. Hamlyn (*The Psychology of Perception* [London: Routledge and Kegan Paul, 1957]) and N. Malcolm ("The Myth of Cognitive Processes and Structures," in *Cognitive Development and Epistemology,* ed. T. Mischel [New York: Academic Press, 1973]) that empirical psychology has little to offer to epistemology.

70. C. Hempel, *Aspects of Scientific Explanation* (New York: Free Press, 1965), p. 127.

71. "What Theories Are Not," p. 241.

72. R. Carnap, "Empiricism, Semantics, and Ontology," *Revue Internationale de Philosophie,* 11 (1950):20–40.

73. For example, W. Sellars, *Science, Perception and Reality* (London: Routledge and Kegan Paul, 1963); R. Burian, "Scientific Realism, Commensurability, and Conceptual Change: A Critique of Paul Feyerabend's Philosophy of Science" (Ph.D. dissertation, University of Pittsburgh, 1971); W. Rottschaefer, "Ordinary Knowledge in the Scientific Realism of Wilfred Sellars" (Ph.D. dissertation, Boston University, 1972); C. Hooker, "Systematic Realism" (ms.); and R. Boyd, "Realism and Scientific Epistemology" (ms.).

ADDITIONAL REFERENCES

Feyerabend, P. "Problems of Empiricism." In *Beyond the Edge of Certainty,* edited by R. Colodny. Englewood Cliffs, N.J.: Prentice-Hall, 1965.

Helmholtz, H. von. *Treatise on Physiological Optics.* Translated from the 3d German edition (1867) by J. Southall. New York: Dover, 1962.

Newton, I. *Mathematical Principles of Natural Philosophy.* Edited by F. Cajori. Berkeley: University of California Press, 1934.

Shimony, A. "Perception from an Evolutionary Point of View." *Journal of Philosophy,* 68 (1971):571–83.

———. "Scientific Inference." In *The Nature and Function of Scientific Theories,* edited by R. Colodny. Pittsburgh: University of Pittsburgh Press, 1970.

ROBERT EFRON, M.D.
University of California, Davis

Biology Without Consciousness—and Its Consequences

> The physiology of the senses is a border land in which the two great divisions of human knowledge, natural and mental science, encroach on one another's domain; in which problems arise which are important for both, and which only the combined labour of both can solve.... For apprehension by the senses supplies after all, directly or indirectly, the material of all human knowledge, or at least the stimulus necessary to develop every inborn faculty of the mind. It supplies the basis for the whole action of man upon the outer world; and if this stage of mental processes is admitted to be the simplest and lowest of its kind, it is none the less important and interesting. For there is little hope that he who does not begin at the beginning of knowledge will ever arrive at its end.
>
> —Hermann L. F. Helmholtz, 1873

1. INTRODUCTION

> I myself find it philosophically quite acceptable, if a friend from the Indian subcontinent tells me that he does not find it difficult to assume that a crystal may be suffused with satisfaction derived from the regularity in the arrangement of its atomic particles. I do not find anything distasteful in this assumption. I am aware, of course, that such experiences.... are unrecognizably different from ours. On the other hand, ... I ... am quite willing to endow monistically constituted *organisms* with mind-like properties.

These words climaxed a speech given in 1964 by the past president of the Association for the Study of Animal Behavior.[1] This imputation of consciousness to inanimate matter is not unusual today. Twentieth-century biologists are making statements that are flatly indistinguishable from the propositions of ancient Indian mysticism. Why?

The science of biology suffers from a progressive and potentially fatal epistemological disorder. It is characterized by such profound chaos in the realm of definitions and the logical relationships between concepts that many biologists appear to have lost cognitive contact with reality. One of the most fundamental causes of this disorder is a philosophical principle: It holds that all the phenomena of life will ultimately be reduced to, that is, accounted for, described by, and deduced from, the laws of physics and chemistry. It is known as the principle of reduction.

Blind adherence to this principle has resulted in an intellectual smash-up in neurophysiology and psychology. It has involved such concepts as perception, emotion, the reflex, abstraction, the conditioned reflex, the cerebral localization of function, learning, and memory—to name only a few. This essay will document the severity of this disorder in neurophysiology and psychology by presenting a detailed analysis of two of these concepts, the reflex and memory.[2]

2. THE PRINCIPLE OF REDUCTION

The most important premise of the philosophy of materialism which affects the field of biology is the "principle of reduction," the premise that all the phenomena of life can be accounted for, described by, and deduced from the laws of physics and chemistry. As usually stated, this principle is of questionable meaning. In the first place, it is ambiguous. It can be interpreted to mean either that the phenomena of biology will, at some future date, be reduced to the laws of physics and chemistry which are now known; or that biology can be reduced to laws of physics and chemistry which will ultimately be discovered but which are at this moment still unknown.

The second and more serious fault found in the usual statement of the principle of reduction is the fact that the terms *physics, chemistry,* and *biology* are left undefined, on the implicit premise that these terms have clearly defined meanings. Unfortunately, this is not the case. Webster's Dictionary defines physics as "a science that deals with matter and energy." Chemistry is defined as "a science that deals with the composition, structure, and properties of substances." Since chemistry, by this definition, is subsumed under the broader concept of "physics," only the term "physics" is needed.

If physics is defined in such broad terms, however, then *all* knowledge of the universe is to be subsumed under the "science of physics." Physics actually becomes that science which studies the nature of reality. In a paper in the *Journal of Philosophy,* one physicist actually said, "In the widest sense, physics is the study of all the phenomena that occur in nature, and its problem is to understand them."[3] According to this view, economics, ethics, and esthetics, as well as biology, are the province of physics. Indeed, if physics is considered to be the study of all phenomena in nature, the study of reality, then it is merely a truism to assert that biology is reducible to the laws of physics. The use of the conventional definition of physics, then, turns the principle of reduction into a fatuous proposition.

This principle becomes intelligible only when the *traditional* definitions of physics and biology are used. Historically, the body of knowledge referred to as "physics" studies the nature and actions of *inanimate* entities; historically, the body of knowledge referred to as "biology" studies the nature and actions of living entities. Based upon these more restrictive and accurate definitions, the principle of reduction asserts that every action of a living entity will be accounted for, described by, and deduced from those laws of physics which are *entirely* derived from a study of *inanimate* entities, that is, that there are no fundamentally different principles of action (causal factors) found in living as contrasted to inanimate entities. Thus formulated, the principle is at least of significance and is, in fact, the implicit definition in use by biologists—and in use throughout the rest of this essay.

How is the principle of reduction validated? The two basic forms in which it is supported divide reductionistic biologists into "hard" and "soft" varieties. The "hard" reductionist claims that the principle of reduction must be accepted on faith.[4] The "soft" reductionist claims that the principle of reduction is not necessarily true, but that we should accept it because of its heuristic value. Let us examine each of these positions in more detail.

The "hard" reductionist is an explicit mystic (mysticism being defined as the acceptance of an idea on faith and without logical evidence). He claims to know in advance that all the phenomena of life (including consciousness) will *necessarily* prove to be reducible to, explicable by, or expressed in terms of the laws of physics derived from the study of inanimate entities. Since we do not know that these phenomena will be reduced to the laws of physics until it is so proved, the "hard" reductionist is literally claiming omniscience. (To grasp the absurdity of this epistemological position, imagine, for example, a physicist who claimed that the laws of gravitational attraction will be fully accounted for, described by, and thus deducible from the laws governing subatomic particles—and those laws alone. While this claim might ultimately be proved correct, it cannot, of course, be accepted as knowledge without entailing a claim to omniscience. Few physicists would be willing to make this claim. The "hard" reductionist, however, does not have this reluctance.) The "hard" reductionistic biologist openly holds his position as an act of faith. K. Lashley, a distinguished biologist, has expressed the reductionist's credo as follows: "Our common meeting ground is the *faith* to which we all subscribe. . . . I believe, that the phenomena of behavior and mind are ultimately describable in the concepts of the mathematical and physical sciences."[5] Such mysticism must be rejected by any rational scientist.

The "soft" reductionist—at least overtly—does not voice the reductionist's credo. He accepts the principle of reduction without logical evidence because he thinks that any alternative to this premise will inevitably lead to the idea of spiritual or mental "substances." He thinks that he must accept either the principle of reduction or "ghosts in the machine." Given this (false) alternative, he adheres to the principle of reduction although he knows that he cannot prove its validity.

The "soft" reductionist believes that it is proper and at times scientifically profitable to accept an idea which is an unproved arbitrary assertion. The biologist, he says, should first attempt to explain the particular phenomenon of life which he is investigating by using the concepts of physics and chemistry. It is only in the event of ultimate failure to explain the phenomenon by this means that he should expend his energies on developing new (nonreductionistic) conceptual frameworks. In his view it is a waste of time to do this thinking now, since he expects that the principle of reduction will ultimately turn out to be true.

The "soft" reductionist does not see that the biologist is unable *in principle* to follow this recommendation. By what epistemological or biological principle can the "soft" reductionist decide, at any given stage of his knowledge, that a particular phenomenon will not eventually (ultimately) turn out to be explicable by the laws of physics? As he has no way (short of omniscience) of knowing when to call it quits, he is forced by his implicit faith in the principle of reduction to adhere to it either until he is proved right or, failing this, until he dies. (A physicist who claims that he accepts, for its heuristic value, the idea that gravitational attraction will ultimately be reduced to the laws governing subatomic particles would face identical difficulties. By what epistemological or physical principle could he decide, at any given stage of his knowledge, that the phenomenon of gravity will *not* eventually be reduced to the laws of subatomic particles? There is no evidence that would ever satisfy him that his premise was incorrect.)

The "soft" reductionist, in essence, is as much a mystic as the "hard" reductionist, who overtly accepts the principle of reduction on faith. The "soft" reductionist has another faith as well: He apparently believes that the biologist will *somehow* know when to abandon the principle of reduction. The argument of "soft" reductionism is as unacceptable as the argument of the "hard" variety, and for the same reasons: It presupposes omniscience and is anchored in faith. Neither the "soft" nor the "hard" reductionist is discouraged by the practical failure to reduce the phenomena of life and of consciousness to the laws of physics. Failures are always accounted for by the inadequate state of existing knowledge. The reductionist has absolute faith that the unexplained phenomena of life and of consciousness will be reduced to the laws of physics *in the future.*

It must be stressed that the issue which concerns me here is *not* the truth or falsehood of the principle of reduction. There is no philosophical or scientific obligation to refute an idea for which no evidence has been adduced and which has the epistemological status of an arbitrary assertion. My focus in this essay will be on the consequences, in the field of biology, of the fact that too many scientists have accepted the principle of reduction on faith.

To accept any idea on faith converts it into a not-to-be-questioned (mystical) dogma which cannot be modified, extended, or discarded as a result of further knowledge. It is the mental equivalent of a strait jacket. In particular, the acceptance of an *explanatory* concept on faith causes disastrous epistemological consequences because it inverts the very purpose of an explanation. The epistemological role of an explanation is to account for some aspect of reality which we do *not* understand on the basis of concepts which *have* already been validated. An explanation based upon arbitrary assertions represents an attempt to account for some aspect of reality by using concepts which *have not* been validated. A rational scientist relies on man's knowledge: He accounts for the unexplained in terms of the *known.* The mystical scientist relies on man's ignorance: He tries to account for the unexplained in terms of the *unknown.* To attempt to "explain" a phenomenon by means of the unknown severs epistemology from reality; the thinking process is not anchored in fact.[6]

Only when a hypothesis is based on fact can it be checked for error. The scientist can discover that he is led to factual contradiction, that additional knowledge is needed, and that his thinking process must be checked. When a hypothesis is not based on fact, the scientist has no means of discovering that an explanation is erroneous; its invalidity cannot be discovered by *him* until a full explanation is discovered by *someone else.* Until that time, the scientist who operates on faith has no motivation or methodology to check his thinking process. He already has his explanation.

The acceptance of an idea on faith inverts the epistemological process in one other way. Instead of forming concepts which correspond to reality, those who accept an idea on faith must now make reality correspond to their not-to-be-questioned dogma. Since reality will not, in fact, accommodate itself to a man's beliefs, such a man distorts his view of existence so that it *appears* to correspond to the idea which he holds on faith.

In sum, the acceptance of *any* idea on faith leads to restriction, stagnation, paralysis, and distortion of the thinking process.

The acceptance of the principle of reduction on faith has led to a particular set of distortions and to ultimate self-contradiction. The reductionistic biologist accepts as an absolute that *no fundamentally different principles of action are found in living as contrasted to inanimate entities.* He sees biology only as the study of the means by which the actions of inanimate entities are combined to produce the phenomena of life. He thus restricts himself to the study of those actions of living entities which are *similar* to the actions of inanimate entities, and he evades focusing attention on those actions of living entities which are significantly *different* from any action of inanimate entities.

This evasion becomes massive when the reductionist attempts to deal with the one phenomenon encountered in living organisms which does not remotely resemble anything found in the inanimate world—the phenomenon of consciousness. The greatest conceptual stagnation and the most marked distortion has surrounded all attempts to reduce consciousness to the laws of physics. It is here that the equivocations and evasions of the reductive materialist are most clearly observed. To be a reductionist is to acknowledge (if only implicitly) that something exists which must be "explained." The reductionist does not usually deny the existence of consciousness. His equivocations become manifest only when he talks of "reducing" the phenomenon of consciousness to the laws of physics.

To reduce consciousness to the laws of physics would require an answer to the following question: "What are the properties, attributes, and functions of consciousness that I am attempting to explain?" A reductionist, in order to succeed, would have to know the "laws of consciousness" before he could know that he had explained them by or deduced them from the "laws of physics." Since the reductionist believes that all "laws of consciousness" are at root merely "laws of physics," he should, if he were consistent, not only grant the existence of consciousness, but actively study the phenomena of consciousness as important or unusual manifestations of matter and energy. Instead, he usually denies their importance or their significance by claiming that the phenomena of consciousness, by their nature, cannot be studied "scientifically." If this excuse were valid, then consciousness could never be "explained" by the laws of physics!

The rationale for this evasion ultimately rests on the reductionist's concept of causality. He holds that the cause of any event is the occurrence of a preceding *physical* event. By physical event, he means, of course, an event describable in terms of the actions of inanimate entities. Consciousness, he maintains, is the end product of a complex sequence of physicochemical events, each of which is fully determined by preceding physicochemical events. Consciousness is an epiphenomenon at the *end* of this causal chain. As a mental phenomenon, it can have no causal efficacy;[7] that is, although it is caused, it cannot be the cause of any further events. It is literally conceived of as a metaphysical dead end. Thus, the reductionists find no reason to use the fact of consciousness as an *explanatory* concept: If a phenomenon can cause nothing, it can explain nothing.[8]

A phenomenon which can cause nothing, and thus has no effects, cannot be measured. Many reductionists thus justify their refusal to study any of the phenomena of consciousness on the grounds that "consciousness cannot be measured." When we examine this assertion carefully, however, we will find that it actually contradicts the principle of reduction.

To understand the literal meaning of the idea that consciousness cannot be "measured," one must first examine the concept of measurement itself. Measurement is the process of ascertaining the magnitude of some particular attribute of an entity or of an entity's action. We measure the attribute of length of an entity in centimeters, the attribute of mass of an entity in grams, or the attribute of charge of an entity in electrostatic units. Similarly, we measure the attribute of heat in calories, or the attribute of work in gram-centimeters when we quantify an *action* or process of an entity. Note that a *different* unit of measure is used for each type of attribute—a unit appropriate to that particular attribute.[9]

When they make the claim that "consciousness cannot be measured," reductionists are actually referring to the fact that no aspect of a conscious experience can be measured in ergs, grams, centimeters, or calories—which is to say that no aspect or attribute of the conscious experience of a living entity can be measured in units appropriate to the attributes of inanimate entities and their actions. Most reductionists do not realize that by making this statement, they are flatly contradicting the principle of reduction: If consciousness cannot be measured in units which are applicable to inanimate entities, it cannot be reduced to those laws of physics which are expressed in these units.[10]

Some reductionistic biologists adopt a different position. They hold that consciousness is identical to—the same as—the physiological and physicochemical actions of the brain. They believe that the reduction of consciousness to the laws of physics will be achieved when every different or unique conscious state or experience is *correlated* with a particular or unique physicochemical state of the brain. This position is referred to as a "psychoneural identity theory."

There are two implications of all psychoneural identity theories which contradict the principle of reduction. In the first place, it is a blatant contradiction to argue that mental states are identical with physicochemical states and at the same time to maintain that the two can be correlated. To correlate is to compare the occurrence of, or the association between, two *different* existents, the causal relationship being unknown or unstated. One cannot, in logic, argue that mental states are identical with physicochemical states and at the same time claim that one can correlate the two. For example, a particular frequency of discharge of neurons in the optic nerve may be correlated with a particular experience of "brightness." The frequency of discharge and the experience of brightness are not, however, identical. If they were identical, there would be nothing to correlate! The psychoneural identity theorist thus faces an insurmountable contradiction; the moment he attempts to establish a correlation he has implicitly acknowledged that the two phenomena are different.

In the second place, those who maintain psychoneural identity theories implicitly concede their inability to describe the *causal* relationship between these physicochemical events in the brain and conscious experiences. To correlate physico-chemical events in the brain and conscious experiences is a scientifically appropriate activity. It cannot, however, be construed to be epistemologically identical with an *explanation,* which is a statement of the causal relationship between the events which have been correlated. Those who argue that a correlation is the only form of knowledge that men can achieve deny the possibility of all explanations (as defined earlier). But to accept this epistemological position necessarily leads to the view that consciousness cannot be *explained by* the laws of physics—which is to deny the validity of the principle of reduction.

Blind faith in the reductionist position has not merely led biologists into blatant self-contradiction. It has led them into a complex manipulation of concepts to camouflage that contradiction. They have adopted an epistemological method which makes it appear that reality does conform to their arbitrary beliefs. This epistemological method is well known: It is called definition-switching.

In principle, the method consists of arbitrarily defining the contradiction away. The reductionist attacks the definition and usage of every word which has historically referred to an action of a living entity: "memory," "reflex," "free will," "cognition," and so forth. He then redefines that same word so that it will be applicable to an action of an inanimate entity. By using this epistemological technique he deludes himself into thinking that inanimate entities have the *same* properties that are found in living organisms, that a common denominator has been found, and that the problem of reduction is "solved." The "solution" is primitive animism expressed in scientific jargon: Rocks have "memory," electrons have "free will," molecules have the capacity to "recognize" each other—and crystals are "satisfied" with their own nature.

These characteristic features of the reductionist's epistemology will now be illustrated by examining the deterioration and ultimate destruction of the concepts of reflex and memory.

3. THE REFLEX

The concept of the "reflex" was first formulated by Descartes in 1649 to give plausibility to his (prior) philosophical notion that the lower animals are merely complex automata—machines devoid of mind and of volition.[11] Descartes conceived of the animal's responses to sensory nerve stimulation as inevitable, machinelike, automatic, involuntary reactions which are determined by the innate structure of its nervous system. The very word *reflex* derives from his idea that the effects of sensory stimulation are "reflected" back out from the brain, in particular from the pineal gland.

More than 275 years later, Pavlov wrote,

> Psychologists have studied and are studying at the present time these numerous machine-like, inevitable reactions of the organism—reflexes existing from the very birth of the animal, and due therefore to the inherent organization of the nervous system.[12]

Thus, from its inception in the middle of the seventeenth century to the beginning of the twentieth century, reflex action was taken to be synonymous with the automatic, machinelike responses of muscles and glands mediated by specific, innately "wired-in," neural connections between receptor and effector organs. From its inception to the beginning of the twentieth century, the reflex has stood in contrast to those other actions of the organism which are dependent upon consciousness and volition.

The definition of "reflex" found in Webster is a precise formulation of the concept of the reflex as it developed historically:

> An act (as a movement) performed automatically and without conscious volition in consequence of a nervous impulse transmitted inward by afferent fibers from a receptor to a nerve center and commonly through adjustor neurones outward by efferent fibers to an effector (as a muscle or gland).

To analyze this definition: The genus of "reflex" is an action performed by a living entity. The differentiae are (a) the action is automatic, that is, performed without conscious volition, and (b) it is a consequence of a stimulus to a sensory receptor that initiates a nervous impulse which is transmitted along a specific neuroanatomical pathway to an effector organ ("reflex arc").

These differentiae imply:

1. That other classes of action exist which are performed *non*automatically, *with* conscious volition. If volitional (*voluntarily initiated*) acts did not exist, there would be no point in using the terms "automatic" or "involuntary" as a criterion to differentiate one class of action from another, for *all* actions would be automatic or involuntary.

2. That classes of action exist which are *not* precipitated by an antecedent stimulus to a sensory receptor. If there were no such actions, the notion of a sensory "trigger" would be superfluous and would be of no value in differentiating one class of action from another, for *all* acts would be triggered by a sensory input.
3. That classes of action exist which *are* fully automatic and involuntary but which are due to mechanisms *other* than those of the reflex arc. If there were no other mechanisms underlying the automatic and involuntary activity of living organisms, there would be no value to the concept of the reflex arc, for *all* automatic activity would be dependent on such arcs.

The definition of "reflex" action contains, therefore, by implication, reference to a class or classes of action which are nonreflexive. This is necessarily so, since the purpose of a definition is to isolate the *essential* attributes which differentiate one phenomenon from another. Behavior which is automatic, innate, involuntary, and independent of consciousness needs to be isolated conceptually only because *other* behavior exists which is voluntary, learned, and dependent on conscious activity.

The concepts of "consciousness" and "volition" logically antecede and are presupposed by the concepts of the "nonconscious" and the "nonvoluntary (automatic)." It is only the prior concepts of "consciousness" and "volition" which make the concept of "reflex" epistemologically possible and meaningful. To attempt to use a concept while at the same time denying the validity of the concepts on which it rests is not logically permissible. In the field of physics, for example, it would be illogical to use the concept of an electron, while repudiating the concepts of charge and of mass. Since an electron is an entity with a mass of 9.1085×10^{-28} grams and a negative charge of 4.80×10^{-10} electrostatic units, the concept "electron" would have no meaning if the concepts of charge and of mass were spurious.[13] The concept of a "reflex" genetically depends upon and presupposes the concepts of "consciousness" and of "voluntarily initiated action." The concept of a "voluntarily initiated action" itself presupposes the concept of "consciousness" and cannot be formed or grasped in the absence of this concept.

For reasons given in section 2, the reductionist cannot use a concept of the reflex, which rests explicitly upon the concepts of volition and consciousness—concepts which he cannot "reduce" to the concepts of physics. He has, therefore, redefined the reflex in such a fashion that it no longer rests upon the "unscientific" concepts of consciousness and volition. He has done this over a period of seventy years, by means of gradually eliminating the differentiae of the definition of the reflex.

In 1863, I. M. Sechenov said, "To come under the category of reflex movements, it is *only* necessary that the movement should clearly result from stimulation of a sensory nerve and be of involuntary nature."[14] Sechenov thus omitted from his definition of the reflex the crucial differentia that the reflex must be independent of the faculty of consciousness. In 1880, R. Heidenhain considered the response of a hypnotized subject to a verbal command to be a "reflex," because he considered this action to be involuntary. He apparently failed to recognize that only conscious men understand verbal instructions.[15] In 1888, C. Richet called the eye-blink response to a threat a "psychical reflex," because there was awareness of the stimulus. He considered habits to be acquired "psychical reflexes."[16] In 1899, Jacques Loeb, the theoretical mechanistic biologist, considered phototropisms in plants, and actions of organisms following removal of their entire central nervous systems, to be reflexes. He said, "A study, then, of comparative physiology brings out the fact that irritability and conductibility are the *only* qualities essential to reflexes, and these are both common qualities of all protoplasm."[17] Loeb thus eliminated *all* previous differentiae from his definition of a reflex—even the necessity of considering a reflex to be a function of the nervous system! In 1931, B. F. Skinner defined the reflex as an "observed correlation of stimulus and response."[18] With this, the definition of the reflex was in effect declared free not only of all differentiae, but even of its genus, which classified the reflex as an action of a living entity.

Observe that once the concept of *life* is withdrawn, the concept of the reflex merely pertains to *any* causal relationship between two events. By Skinner's definition, the observed correlation of the response of a photocell and the arrival of a photon, and a man's response to a woman in the act of falling in love, would both be "reflexes." This is the inevitable consequence of removing the genus and the differentiae of a definition.

While the physiologists were progressively abandoning the differentiae of the definition of the reflex, they were—necessarily—becoming more confused as to what a reflex really was. Many began to wonder if a reflex even existed at all. Thus, Sir Charles Sherrington maintained that a "pure" or "simple" reflex did not actually exist. He concluded that the "simple reflex is a convenient, if not a probable fiction" because he could not define the reflex in purely physiological or anatomical terms.[19] This is to be expected, for it is neither a physiologically definable mechanism nor an anatomically definable structure. *The reflex is definable only by reference to the concept of voluntarily initiated action.*

The physiological and anatomical findings, regardless of how detailed they may be, can never be used as a means of escaping the fact that the concept of the reflex is hierarchically dependent upon the concept of consciousness. This will be true even when the *physiological* difference between a reflex action and a voluntarily initiated action has been *fully* specified. To make this point clear, let us assume, for example, that in the year 2000, we find that a particular, fully specifiable, electrical potential recorded from a fully specifiable region of the brain is *always* found when a reflex action occurs and is *never* found with a voluntarily initiated action. The physiologist who makes this elegant discovery might conceivably say, "We will hereafter *define* the reflex as an action which is perfectly correlated with this potential. We no longer need the concepts of consciousness or of voluntarily initiated action in our scientific vocabulary." Even after such a discovery, this would be an error: The physiological facts of this correlation could not have been discovered without *using* the previous concepts of the reflex, and thus the concepts of consciousness and of voluntarily initiated movement. The new definition of the reflex still rests upon these concepts and would be meaningless if they were denied.

When modern neurophysiologists fail to grasp or refuse to acknowledge that the reflex must be defined with reference to consciousness and volition, they are driven—necessarily—to one of two conclusions: Either that the reflex does not exist at all, or that every action in response to a stimulus is a reflex. Thus, D. P. C. Lloyd, in 1960, while describing the properties of spinal reflexes, stated that reflexes are "artificial entities."[20] And E. Bykov, a follower of Pavlov's, in the same year said, "All actions of the organism in response to stimuluation of the receptors and involving the central nervous system are reflexes."[21]

If the reflex is an "artificial entity," "an improbable fiction," a synonym for any neural response to a stimulus, or, as Skinner claims, a "gratuitous" assumption, why is the term in such constant use by neurophysiologists, psychologists, and animal trainers? The answer should now be clear. The reductionistic biologist retains and uses the word "reflex" because it enables him to make *implicit* use of the *old* concept of the reflex (involuntary behavior independent of consciousness) without admitting that his new definition still logically rests upon the concepts of consciousness and volition. He needs the concepts of "automatic" and "involuntary," but wishes to evade the fact that the use of these terms is meaningful only by virtue of the existence of nonautomatic and voluntary actions. This epistemological procedure is known, in some scientific circles, as "having your consciousness and eating it, too."

Let us now take another concrete illustration of the epistemological deterioration in the field of biology caused by adherence to the principle of reduction. When we turn to the concept of memory, the technique and the destructive consequences of definition-switching will be even more obvious.

4. THE CONCEPT OF MEMORY

The experiences of recall, recognition, familiarity, and reminiscence are known to all men by introspection, and have been acknowledged as phenomena of consciousness for more than two thousand years. Aristotle defined memory as "neither perception nor conception, but a state or affection of one of these, conditioned by lapse of time."[22] He held that whenever one exercises the faculty of remembering, one says within himself some equivalent of "I formerly heard this" or "I formerly had this thought." Some two thousand years later, William James defined memory as "the knowledge of a former state of mind after it has already once dropped from consciousness; or rather it is the knowledge of an event or fact, of which meantime we have not been thinking, *with the additional consciousness that we have thought or experienced it before.*"[23]

The term "memory" includes the retention, recall, and recognition of facts whether in perceptual or conceptual form. Although philosophers have long debated the relationship between memory and images, memory and knowledge, memory and time, we need not discuss these issues in this essay. All that is necessary for our present purposes is to recognize that the concept "memory" *depends upon and presupposes the concept of consciousness,* cannot be formed or grasped in the absence of this concept, and represents, within wider or narrower limits, a specific type or state of conscious activity.[24]

Just as the definitions of the reflex became less and less specific as the nineteenth century ended, so did the definitions of memory. The process of disintegration was accomplished by identical means—the elimination of crucial differentiae which had isolated the concept of memory from other concepts. It took only fifty years for experimental psychologists to destroy the concept of memory.

The base was laid by Hermann Ebbinghaus. In 1885, in a monograph entitled *On Memory,* Ebbinghaus identified three types of memory, two of which had been long recognized: The first is the type of memory which can be "called back into consciousness by an exertion of the will"; the second is the type of memory which pops into our minds "spontaneously"—that is to say, without an effort of will. The third was original with Ebbinghaus and contained a crucial conceptual error. He claimed, without offering any logical reason, that there was a form of memory in which the "vanished mental states" may never return to consciousness at all, but yet "give indubitable proof of their continuing existence. . . . Most of these experiences remain concealed from consciousness and yet produce an effect which is significant and which authenticates their previous existence." The existence of this type of memory, he said, could only be deduced from the effects it produces.[25]

Ebbinghaus claimed that he had developed a method of measuring these effects which has since become known as "the method of savings." It consists of asking a person to commit some type of verbal material (usually nonsense syllables) to memory and counting the number of practice trials required to achieve perfect repetition. At some later time, when the person can no longer repeat this verbal material, he is again asked to memorize it. The number of trials required to achieve perfect repetition is found to be less than was required on the first occasion. The amount of "savings," as measured by the arithmetical difference between the two scores, was considered by Ebbinghaus to be a "measure of memory." As the "savings" could be numerically measured with a high degree of accuracy, he declared that memory could, for the first time, be quantified. This quantitative method of studying the phenomenon of memory was hailed as a triumph of scientific psychology.

Observe that Ebbinghaus retained the genus of the classical definition of memory (a capacity of a living entity) but eliminated the differentiae (a conscious recall or recognition of a previous experience). His concept of memory thus ceased to refer to a conscious experience, but now referred to something which *resulted* from a conscious experience and which had the property or capacity to alter subsequent performance (behavior).

While for Ebbinghaus, the original genesis of the "something" was a conscious experience, those who followed him historically dropped all reference to consciousness. After Ebbinghaus, memory was considered by most psychologists and neurophysiologists simply to be *something in a living organism which caused or which made possible an alteration of behavior.*

In the twentieth century, it was found that Ebbinghaus's methodology was applicable not merely to the memorization of lists of nonsense syllables, but to all sorts of other activities in men and in lower organisms. It was realized that the method of savings, which had been considered by Ebbinghaus to be a specific measure of "type III memory," also measured something which psychologists were calling "learning." There was a choice to be made: Either the method of savings had to be abandoned as a nonspecific measuring technique, or the concepts of memory and learning had to be fused.

The choice was to fuse them. It was inevitable, given the poorly defined state of the concept of "learning." The term "learning" had been used to refer to the acquisition of perceptual discriminations, acquisition of motor skills or habits, problem solving, adaptation, association, "insight" solution, secondhand concept acquisition, and at times the discovery of new conceptual knowledge. Thus we say, "the child *learns* to walk" (motor skill), "the child *learns* the difference between a red and blue ball" (discrimination). "The worm *learns* how to get through the maze" (problem solving?). "The schoolboy *learns* that light travels faster than sound" (secondhand concept acquisition). "Pythagoras *learned* that the sums of the squares erected on the hypoteneuse . . ." (discovery of new knowledge).

"Learning," therefore, was generally used to refer to the acquisition of motor skills and of perceptual and conceptual knowledge. It did not *exclude* the concept of consciousness, it rested on and presupposed it; but it stressed the one common feature of all these activities of living entities—the acquisition of a new potentiality for action. For twentieth-century psychologists, learning, practically speaking, referred only to the alteration of behavior or of performance as a result of previous experience. Since memory, after Ebbinghaus, was considered to be that which made the alteration of behavior (performance) possible, the fusion of these two views was almost inescapable: *Memory was now considered to be something that results from previous experience which causes learning or makes it possible.*

The link between the concept of memory and the concept of consciousness was explicitly broken when "learning," like all other concepts of mental functions, was redefined in the twentieth century. Once the concept of consciousness was eliminated from the concept of "learning," it necessarily vanished from the concept of memory. Since the 1930s almost all discussions of memory to be found in the standard textbooks of psychology and physiology militantly disregard consciousness and are restricted to the question: "What is the physical nature of the something which makes learning possible?"

A typical answer to this question always includes reference to a presumed physical change in the tissues of the organism. For example, in the 1938 edition of R. S. Woodworth's textbook of experimental psychology, he writes:

> In learning, work is done by the organism; this work leaves after effects which we may include under the noncommital term, *trace*. What is retained is this trace. The trace is a modification of the organism which is not directly observed but is inferred from the facts of recall and recognition.[26]

For Woodworth, the cause of memory *is* the physical trace. It is this physical trace, or engram, which makes learning possible.

The more recent behaviorists object to Woodworth's discussion because of his reliance on the facts of "recall" and "recognition"—facts referring to conscious experience. The behaviorists assert that science should deal only with "measurables" and declare that all we can observe or measure is the behavior, or response tendency, of an organism.[27] They "purify" such formulations as Woodworth's, eliminate every implicit reference to consciousness, and conclude that memory is *that physical change in the tissues of the organism which alters its subsequent behavior or response tendency.*

In accordance with this view, two authors writing on memory formation in the goldfish state: "The term 'memory' is used to describe the alterations in the brain which are responsible for the increased responding" (of the goldfish to a stimulus).[28] Observe that this definition of memory is virtually identical to Ebbinghaus's concept: Ebbinghaus defined memory as the something which is responsible for the altered behavior. By the 1950s, this was explicitly equated with a physical change, usually located in the brain. The physical change is often located, by reductionists, in other tissues as well. We are told, for example, that there are many different types of *biological* memory: There is genetic memory, immunological memory, instinctual memory, acquired memory, and psychic memory. Genetic "memory," it is claimed, is stored in the DNA molecules of the chromosome. One leading exponent of molecular biology declared that DNA is "the genetic memory of the species."[29] Immunological "memory," it is claimed, is stored in the reticulo-endothelial system. An immunologist has suggested that "the immunologic engram would seem to be stored in a variety of lymphoid cells that one calls immunologically competent cells."[30] Some immunological "memories" are said to be innate (the so-called natural antibodies); others are said to be acquired by exposure to antigens.

We are also told that memory is not a phenomenon seen only in living organisms; there is *nonbiological* "memory" as well. Even the footprint in the sand is considered by some writers to be a form of memory. The hysteresis effect seen in iron is called "memory," and this is the basis of the so-called memory of many computers.[31] A watch which was brought near a magnet and now runs slowly is said to have a "memory" of its exposure to the magnetic field which has altered its subsequent behavior. In a recent paper, a particular chemical reaction involving cupric nitrate and O-phenenthroline was considered to have "memory" because "the system 'remembers' its initial conditions."[32] At an international conference on memory in 1963, one speaker maintained that linseed oil has "memory," because "if linseed oil is exposed to oxygen for a period of time and then put away for 10 years, its oxidative rate when returned to the air will be proportional to how long it had been originally exposed."[33] In an article in *Scientific American,* it was pointed out that a pebble rubbed smooth in a stream bed rolls (reacts) differently as a result of this experience. It therefore exhibits the fundamental characteristic of "memory." The biologist who wrote this article merely considered that the "memory" of the pebble was "uninteresting."[34] Not only do chemical or physical *reactions* have memory, but even individual molecules have memory according to M. Eigen, who has said, "DNA, for instance, has memory."[35]

These examples, and there are thousands of others, should suffice to indicate that the degeneration of the concept of memory has reached its final stage. The differentiae pertaining to consciousness are gone; now the genus—an action of a *living* entity—is gone. The concept of memory, today, refers only to the fact that entities can change their mode of action when one or more of their properties are altered.

Having finally rendered life "irrelevant" to memory, reductionists now consider memory to be a *metaphysical* process, that is, a process exhibited by every existent in the universe, simply by virtue of being an entity with a capacity for change. For the reductionist, there is no longer a difference *in principle* between the so-called memory of sandy beaches, chunks of iron, chemical reactions, pebbles, and DNA molecules, and the radically different phenomenon experienced by men. There is, even for the reductionist who is most skilled at evasion, an obvious empirical difference between the "memory" of a pebble or a chunk of iron and the memory of a man. He needs to account for this difference without violating his basic absolute, that is, without introducing any new principle of action not found in inanimate entities. His only solution is to claim that while *all* material entities have "memory," *some* entities have this metaphysical characteristic in more complex or extensive forms than do others.

The blind adherence to the principle of reduction has led to virtually identical results in such different fields as neurophysiology and psychology. The parallelism of the epistemological destruction is impressive: Just as the concept of the reflex became a "correlation" between a stimulus and a response, so the essence of memory becomes the capacity of an entity to change. Just as for many neurophysiologists, the reflex can be understood only by unraveling the physiochemical properties of the cell which give rise to irritability, so the contemporary student of memory considers that his exclusive problem is to understand the physiochemical properties of the cell which allow it to change its structure or function. Just as the definition of the reflex was dismembered by reductionists so that the genus of life vanished and the concept became identical with a causal—metaphysical—relationship, so was the definition of memory similarly dismembered by reductionists so that the genus of life vanished, and the concept became identical with a metaphysical process or capacity.[36] Just as the reductionists' failure to grasp that the reflex can be defined only with reference to the concepts of life, consciousness, and volition led to such absurdities as the claim that chemotaxis in bacteria is a reflex—so their failure to grasp that memory can be defined only by reference to the concepts of life and consciousness has led to an even more ludicrous position, the widespread idea that molecules have memory.

The most important epistemological similarity of all—and the one which generates all the others—is the attempt to cash in on the established meaning of a concept, while destroying the conceptual base upon which it rests. You will recall that I said that "the reductionist biologist retains and uses the word 'reflex' because it enables him to make *implicit* use of the *old* concept of the reflex (involuntary behavior independent of consciousness) without admitting that his new definition still logically rests upon the concepts of consciousness and volition." Observe now that the reductionist retains and uses the word "memory" because it enables him to make implicit use of the old concept of memory (which refers to such phenomena of consciousness as recognition, recall, reminiscence, and so forth). He cashes in on the fact that *we* will think he is saying something about these phenomena of consciousness when he speaks of the "memory" of the molecule. It is an error for us to think so. The memory of a molecule, as the reductionist is actually using this term, *refers only to the fact that some aspect of an entity can change.* He has substituted for the concept "memory" the concept "change," without admitting it, but continues to exploit the established meaning of memory. If he were not counting on its established meaning, he would be the first to announce that he is using the word "memory" as a synonym for "change." By not using the well-established concept of change—by his insistence on using the word "memory"—he is able to exploit the fact that for two thousand years, this concept has referred to phenomena of consciousness. In the very last analysis, the reductionist is *counting* on this implicit reference to consciousness in order to attack the scientific legitimacy of the concept of consciousness, and to "solve" the problem of reductionism.

5. CONCLUSION

The reductionist has attempted to solve all his problems by dissolving definitions. He has attempted to "explain" the phenomena of life by obliterating the distinction between living and nonliving entities. He has attempted to "explain" the phenomena of consciousness by obliterating the distinction between conscious and nonconscious entities. The confusions which this pretense at an explanation has generated are not restricted to the concepts of the reflex and of memory. The epistemological distortion has involved, to a greater or lesser degree, contemporary thinking about *every* aspect of a living entity which would differentiate it from an inanimate entity, and every aspect of the faculty of consciousness which would differentiate a conscious from a nonconscious entity. To cite a few examples in other realms:

One prominent biochemist says:

> Many types of biochemical memory depend upon *recognition* of one molecule by another. For example, with marked specificity enzymes *recognize* substrates, antibodies interact with antigens, cells *recognize* other cells and so forth. The basis of genetic *memory* is similar, for it depends upon complementary *recognition* of one nucleotide by another via highly specific hydrogen-bonding.

Later he writes that "the question of how cells *perceive* certain stimuli might be pursued on the molecular level by studying chemotaxis in bacteria."[37] Another author refers to the "molecular '*intelligence* system' for distinguishing like from unlike."[38]

In short, throughout the biological sciences, such phenomena as *recognition, identification, perception, intelligence,* and *memory* are now treated as forms of specific chemical reactivity (affinity), rather than as forms of cognitive processes. But observe: These cognitive processes are the means by which we achieve knowledge. Since the reductionist feels compelled to deal with every capacity of a conscious organism as though it were a property of an inanimate entity, he adopts a similar policy toward knowledge itself. Knowledge, too, must be reducible. He has laid the groundwork for this ultimate conclusion: Memory, he says, is molecular change; memory and learning are fused; learning is the means of gaining knowledge—therefore, molecular change is equivalent to a gain or loss of knowledge.[39]

And thus we see how the reductionist could conceive that knowledge might be successfully transferred from one animal to another by transferring molecules from the first to the second. Some of these scientists have posited that this transfer of knowledge might be accomplished by *feeding* an animal which has learned some particular task to an untrained animal, or by grinding up the brain of one animal and *injecting* it into another. Such experiments have actually been performed in several American laboratories, and we were told via all the communications media that they were "successful."[40] Seventy years of gradual disintegration of definitions, blind faith, and evasion have led to a "revolutionary" conclusion that would not startle a savage. The savage has long been quite certain that by eating his enemy he acquires some of his powers. When twentieth-century scientists reduce their epistemological functioning to the level of a savage's, it is not surprising that both arrive at the same conclusions.

It should be mentioned in passing that many of these experiments in both worms and mice cannot be replicated in laboratories other than those in which they were first described, and there are now a number of published articles to this effect.[41] However, the authors of these dissenting papers do not give any indication that they reject the prevailing concept of memory.

Representative of the views of the *dissenters* is a letter to *Science* signed by twenty-three scientists who, in seven different laboratories, had been unable to confirm any of the claims for "transfer of learning by injection of brain-extract from trained donors."[42] Their implicit acceptance of the prevailing views of memory and learning is illustrated in their concluding sentence: "Furthermore, we feel that it would be unfortunate if these negative findings were to be taken as a signal for abandoning the pursuit of a result of enormous potential significance." What is of such "enormous potential significance" in this class of nonreplicable experiments? I quote from the author of the original study on worms:

> Yet, if learning has a biochemical basis, if the engram is really just an alteration in an RNA molecule inside some cell, then it should eventually be possible to specify precisely which chemical changes are associated with which behavioral changes. Once the chemistry is known we should be able to form engrams in a

wide variety of ways. It is very likely that we could synthesize the necessary chemicals in a test tube and then inject the engram ("knowledge") directly into an animal's cells, just as we apparently can now "feed" engrams to cannibal flatworms. Once the "knowledge" became functional (inside the animal), the organism would show "a relatively permanent behavioral change" without ever having directly experienced any contiguity whatsoever between stimuli and/or responses. From merely observing the animal's behavior after the engram became active, we could not, for the life of us, tell whether the animal had "learning in the regular fashion" or whether it had acquired the engram in a much more direct manner.[43]

In other words, what is of "enormous potential significance" in these experiments on memory transfer is the idea that *knowledge* can be injected into animals—that *knowledge* as well as memory may finally be reduced to a *chemical!*

The ideas and speculations relating to memory to which I have been referring are those of the theoreticians. But ideas, whether they are right or wrong, soon have practical consequences. Within a few years of the initial speculation about the role of DNA or RNA in memory, some physicians considered it worthwhile to administer a drug that may increase the biochemical synthesis of RNA to a group of elderly patients with memory defects.[44] This drug has been used as a stimulant in Europe for some years. Although there were side effects of "overstimulation," these investigators argued that the stimulant effect was too slight to be the cause of the apparent beneficial effect on memory. The drug is also reported in use by another group of investigators in an attempt to improve the memories of normal students at a large university. In two recent papers, however, these reported improvements in mental or physical functioning could not be confirmed.[45]

In another study, two psychiatrists administered intravenous injections of an impure RNA extract, obtained from yeast cells, to eighty-four elderly patients suffering from memory defects due to brain disease. The intravenous injections had to be abandoned because they were "apt to produce severe shock-like reactions." Oral administration, which produced only gastrointestinal disturbances, was reported to improve the "memory retention" of the patients.[46] This work has not yet been confirmed.

A scientist interested in computers would not conceive of the idea that he could chop up a computer into its electronic components and then sprinkle these resistors, capacitors, and diodes at random into another computer in the hope that the second computer would perform its calculations faster or with fewer errors. It is absolutely incredible that a biologist could conceive that memory might be transferred in animals or man by such a technique. It is incredible, at least, if one does not fully recognize the epistemological collapse which has occurred in the field of biology—a collapse which can be attributed in large part to the fact that too many biologists continue to adhere to the idea that the concept of consciousness has no scientific or epistemological status, that consciousness is an irrelevant epiphenomenon.

Clearly, the issues discussed in this essay are not only of academic interest. Philosophy and epistemology are not academic disciplines having no practical importance: Human lives are at stake. Indeed, it is the philosopher who is in large part responsible for this intellectual smash-up in biology. He has consistently advocated the use of invidious epistemological remedies for philosophical and scientific problems. He has maintained, at different times and in different places, that there is no reality to know; or that there is one, but we cannot know it; or that there are two, but we can know only one; or that there may be many, but we can know none. He has often maintained that all human knowledge starts with an act of faith. He has, on different occasions, persuaded the biologist that it is impossible to prove a proposition true, but that it is possible to falsify it; or that it is possible neither to verify nor to falsify—that is to say, that proof is not possible at all. He has maintained that we do not really know what we mean when we say that a proposition is true; or that we have defined our terms in such a fashion that the proposition, if true, is so by arbitrary convention. He has maintained that all we can know is what our linguistic conventions permit, and that there is a realm of knowledge outside the realm of these conventions which is beyond the powers of any human being to grasp, except apparently the philosopher who makes this claim. He has also maintained that certainty is impossible, but claims to know this by means which he does not reveal. He has maintained that definitions are arbitrary, or operational, or conventional. He teaches, and appears to believe, that most problems are not problems at all if you merely switch enough definitions. He holds the view that definitions are merely a "matter of convenience" and need not correspond to reality.

Considering the fact that these are the ideas which have dominated the philosophical scene, it is not surprising that the biologist can find a sanction and precedent for *any* arbitrary epistemological act he sees as "convenient" at the moment. Considering the state of its epistemology today, it is not surprising that the science of biology is in chaos.

NOTES

A version of this paper has appeared in the journal, *Perspectives in Biology and Medicine.*

1. O. Lowenstein, "Descartes, Mechanistic Biology and Animal Behavior," in *Learning and Associated Phenomena in Invertebrates,* ed. W. H. Thorpe and D. Davenport. Proceedings

of a Joint Conference of the Association for the Study of Animal Behavior and the Section of Animal Behavior of the American Society of Zoologists (London: Bailliere, Tindall and Cassell, 1964), p. 111, italics added.

2. For a presentation of the epistemological methods underlying this analysis, the reader is referred to Ayn Rand, "Introduction to Objectivist Epistemology," *The Objectivist,* 5, nos. 7–12; 6, nos. 1 and 2 (1966–1967).

3. G. Feinberg, "Physics and the Thales Problem," *Journal of Philosophy,* 63 (1966):5–17.

4. In thus characterizing the position of "hard" reductionists, I am describing the dominant viewpoint of contemporary *biologists.* There are a comparatively small number of *philosophers* today who would argue that reductionism is an *a priori* necessity. This essay is not intended to deal with the metaphysical arguments in support of this philosophical position, but rather with the epistemological consequences of the fact that most practicing biologists actually accept the principle of reduction on faith.

5. K. Lashley, "The Problem of Serial Order in Behavior," in *Cerebral Mechanisms in Behavior,* Hixon Symposium (London: Chapman Hall, 1951), pp. 506–28. Italics added.

6. In any rational employment of the hypothetico-deductive method, the proposed hypothesis which is to constitute the explanation cannot be an arbitrary assumption wantonly asserted out of caprice. The proposed hypothesis must be based upon valid evidence which justifies its serious consideration.

7. This is the root premise of all materialists.

8. This is the explanation for the fact that most reductionistic biologists do not consider psychology to be a science. Since mental events have no effects, the science of the "psyche" must, according to them, be fundamentally invalid. This is the explicit view of behaviorists and psychophysical parallelists.

9. The attributes of consciousness, by their nature, will thus require different kinds of units (standards) of measurement—units appropriate to the attribute being measured. So long as reductionists restrict themselves to studying only those actions of living entities which can be measured in terms of units applicable to inanimate entities, *they can never learn to measure those aspects of biological activity which reflect actions not encountered in inanimate entities.* To claim that "consciousness cannot be measured" is to say either that by its nature it has no attributes which have magnitude or that it does have attributes which have magnitude but that we are unable to measure them. Both of these positions are easily refutable. *Pain,* for example, is an attribute of certain conscious experiences, and it unquestionably has magnitude. *Effort,* to give another example, is an attribute of other conscious experiences, and it too has magnitude. We measure (by means of comparison) the relative magnitudes of many aspects of conscious experience every day. Most reductionists claim, however, that these comparative or relative measurements are "crude," "subjective," not open to "public" verification, and therefore unsuitable as a basis for a science. To adopt this view is to fail to recognize that every science starts chronologically with these comparative measurements. For example, an "erg," one of the most fundamental units of physics, now defined as the work done by "a force of one dyne acting through a distance of one centimeter" arose from the "subjective experience" of *work* or *effort.* If the early physicists had adopted the epistemology of many modern psychologists, and systematically avoided dealing with the experience of effort, because it was "crude" and "subjective," and could not be quantified *initially,* there would be no branch of knowledge called physics today! It is certainly true that, with respect to measurement of psychological processes, we are only at the first phase, the stage of comparative measurement. We have no means of quantifying these measurements more precisely as multiples or submultiples of some specified unit (standard). But this difficulty should be experienced as a challenge to be overcome rather than as an excuse for abandoning the study of consciousness.

10. A. Pap, *An Introduction to the Philosophy of Science* (Glencoe, Ill.: Free Press, 1962), p. 362. The principle involved here was stated by Pap in the following form: "In order for

such a reduction of laws to be possible, however, the concepts of biology must be definable in terms of the concepts of physics and chemistry, since there cannot occur in the conclusion of a valid explanatory deduction concepts that do not occur in the premises or defined in terms of concepts occurring in the premises."

11. A part of this section has previously appeared in my article, R. Efron, "The Conditioned Reflex: A Meaningless Concept," *Perspectives in Biology and Medicine,* 9 (1966):488–514, and is reprinted here with the permission of the University of Chicago Press.

12. I. P. Pavlov, *Conditioned Reflexes* (Oxford: Oxford University Press, 1927), p. 8.

13. I am indebted to Ayn Rand and Nathaniel Branden for identifying and drawing to my attention the wider epistemological fallacy of ignoring the genetic roots of concepts. See N. Branden, *The Objectivist Newsletter,* 2 (1963):2.

14. I. M. Sechenov, *Selected Physiological and Psychological Works of I. M. Sechenov* (Moscow: Foreign Languages, 1952), p. 37. Italics added.

15. R. Heidenhain (1880), quoted in F. Fearing, *Reflex Action* (New York: Hafner, 1964), p. 215.

16. C. Richet (1888), quoted in Fearing, *Reflex Action,* pp. 237–38.

17. J. Loeb (1899), quoted in D. Fleming, ed., *The Mechanistic Conception of Life* (Cambridge: Harvard University Press, 1964), p. 67. Italics added.

18. B. F. Skinner, "The Concept of the Reflex in the Description of Behavior," *Journal of General Psychology,* 5 (1931):427–58.

19. C. Sherrington, *The Integrative Action of the Nervous System* (Cambridge: Cambridge University Press, 1952), p. 7.

20. D. P. C. Lloyd, "Spinal Mechanisms Involved in Somatic Activities," in *Handbook of Physiology,* ed. J. Field (Washington, D.C.: American Physiological Society, 1960), p. 929.

21. E. Bykov, ed., *Textbook of Physiology* (Moscow: Foreign Languages, 1960), p. 523.

22. Aristotle, *De Memoria et Reminiscentia, The Basic Works of Aristotle,* trans. W. D. Ross (New York: Random House, 1941).

23. W. James, *Principles of Psychology* (1890; New York: Dover Publications, 1950), p. 648.

24. W. W. Rozeboom, "The Concept of 'Memory,'" *The Psychological Record,* 15 (1965):329–68.

25. Hermann Ebbinghaus (1895), *Memory,* trans. H. A. Ruger and C. E. Bussenius (New York: Teachers College, Columbia University, 1913).

26. R. S. Woodworth, *Experimental Psychology* (New York: Henry Holt and Co., 1938).

27. The concept of "behavior" or "response" is applicable to both living and nonliving entities, and to conscious as well as nonconscious entities. It is permissible in some contexts, for example, to speak of the behavior (meaning the actions) or the response (meaning the reactions) of a piece of iron in a magnetic field. When we use this terminology, we refer to the physical changes which take place in the iron as a consequence of its exposure to a magnetic field which alter its subsequent reactions. By using general terms such as behavior or response, the reductionist aims at making the presence or absence of consciousness in the iron *irrelevant* to his description. Similarly, by using the general terms behavior or response, the reductionist aims at making the presence or absence of consciousness in the organism *irrelevant* to his science of biology. The implication that consciousness is an irrelevant consideration is in precise accord with the reductionist's explicit epiphenomenalism. By using the words "behavior" and "response," the reductionist considers that he has successfully avoided using the concepts of life and of consciousness.

28. R. E. Davis and B. W. Agranoff, "Stages of Memory Formation in Goldfish: Evidence for an Environmental Trigger," *Proceedings, National Academy of Science,* 55 (1966):555–59.

29. F. O. Schmitt, "The Biomedical Communication Problem," *The Technology Review,* 66 (1964).

30. A. M. Silverstein, "Immunologic and Psychic Memory," *Neurosciences Research Program Bulletin,* 1, no. 1 (1963): Quoted with author's permission, p. 5.

31. It is not irrelevant to point out that Ebbinghaus's method of savings as a means of measuring "memory" is applicable in this situation. Thus, we can infer how magnetized a piece of iron is (how much "memory" it has of its previous exposure to a magnetic field) by measuring how much more magnetic energy we must apply to put it into a state of full magnetization. Ebbinghaus's method is as applicable to a chunk of iron as it is to a human mind—and this is precisely what is *wrong* with it as a means of measuring memory. It is only one particular way of demonstrating that some previous actions or reactions of an entity may modify its subsequent actions or reactions. It does not measure memory at all. Ebbinghaus's error was to equate what is retained with what is remembered.

32. W. Brackman, "A Chemical Reaction with a Memory Effect," *Nature,* 211 (1966), 818–20.

33. D. P. Kimble, ed., *The Anatomy of Memory,* Proceedings, First Conference on Learning, Remembering and Forgetting (Palo Alto, Calif.: Science and Behavior Books, Inc., 1965), p. 9.

34. Ralph W. Gerard, "What is Memory?" *Scientific American,* 189 (1953), 118–26.

35. M. Eigen, "Chemical Means of Information Storage and Readout in Biological Systems," *Neurosciences Research Program Bulletin,* 2, no. 3 (May–June 1964):11–22.

36. G. Bateson, "Exchange of Information about Patterns of Human Behavior," in *Information Storage and Neural Control,* ed. W. S. Fields and W. Abbott (Springfield, Ill.: Charles C Thomas, 1963). Another example of the recent trend toward "metaphysical" definitions is Bateson's concept of learning: "Let us assume that all receipt of information is 'learning.' This will bring within a single theoretical spectrum the whole range of phenomena, beginning with the receipt of a pip by a receiving machine at the end of a wire, up to and including such complex phenomena as the development of neurosis and psychosis under environmental stress." Since every entity in the universe "receives information," learning is, according to Bateson, a metaphysical process.

37. M. W. Nirenberg, "Nucleic Acids in Relation to the Coding of Genetic Information," in *Brain Function: RNA and Brain Function in Memory and Learning,* ed. M. A. B. Brazier (Berkeley and Los Angeles: University of California Press, 1964), pp. 5, 19. Italics added.

38. F. O. Schmitt, "Molecules and Memory," *New Scientist,* 23 (1964):643–45. Italics added.

39. J. V. McConnell, "Cannibals, Chemicals and Contiguity," in *Learning and Associated Phenomena in Invertebrates,* ed. W. H. Thorpe and D. Davenport. Proceedings of a Joint Conference of the Association for the Study of Animal Behavior and the Section of Animal Behavior of the American Society of Zoologists (London: Bailliere, Tindall and Cassell, 1964), p. 65. In one of the more explicit expressions of this conclusion, McConnell stated, "I would like to propose that we re-define learning as being the end product of *any* set of events which causes a (someday hopefully specifiable) change in one or more (RNA?) molecules in an organism's cell(s). Whatever causes the chemical change also causes learning. If no chemical change took place, no learning would take place. We would throw out all of those jargonish intervening variables—such as drive reduction, goal gradients, frustration and incentive, the hypothesis-formations and cognitive maps—that have served a noble purpose in the past but now are obsolete.... At last the hypothetical construct 'learning' could be given a meaning anchored in fact rather than in the never-never jargon of intervening variability." The implicit "scientific" premise underlying this view is that the only valid meaning of "learning" is one in which "learning" is unequivocally defined in terms of molecular change, that is, in terms of physics and chemistry. The implicit epistemological premise underlying this mode of definition is the view that men may define

their concepts in terms of what they *do not know* (their unvalidated guesses, hopes, or faith) rather than in terms of what they *do know* about the existents subsumed by their concepts.

40. J. V. McConnell, "Memory Transfer through Cannibalism in Planarians," *Journal of Neuropsychiatry,* suppl. 1, 3 (1962):42–48.

41. A. L. Hartry, P. Keith-Lee, and W. D. Morton, "Planaria: Memory Transfer through Cannibalism Re-examined," *Science,* 146 (1964):274–75; M. Luttges, T. Johnson, C. Buck, J. Holland, and J. McGaugh, "An Examination of 'Transfer of Learning' by Nucleic Acid," *Science,* 151 (1966):834–37; C. G. Gross and F. M. Carey, "Transfer of Learned Response by RNA Injection: Failure of Attempts to Replicate," *Science,* 150 (1965):1749; and H. M. Brown, R. E. Dustman, and E. C. Beck, "Sensitization in Planaria," *Physiology and Behavior,* 1 (1966):305–08.

42. Byrne, Samuel, Bennett, Rosenzweig, Wasserman, Wagner, Gardner, Galambos, Berger, Margules, Fenichel, Stein, Corson, Enesco, Chorover, Holt, Schiller, Chiapetta, Jarvik, Leaf, Dutcher, Horovitz, Carlson, "Memory Transfer," *Science,* 153 (1966):658.

43. J. V. McConnell, *The Anatomy of Memory,* ed. in D. P. Kimble (Palo Alto, Calif.: Science and Behavior Books, Inc., 1965), p. 310. Even if DNA or RNA, or some other alteration of structure is demonstrated, at some future date, to be the physical means by which "information is stored," this will not account for memory. (See W. Dingman and M. B. Sporn, "Molecular Theories of Memory," *Science,* 144 [1964]:26–29). We would still require an explanation for the means by which this mechanism makes possible a *conscious* experience of a similarity between some *aspect* of previous perception, experience, or thought and some aspect of a current perception experience or thought. We would still require an explanation for the means by which we abstract and store the meanings of our sensory experiences. (See W. W. Rozeboom, "Concept of 'Memory.'")

44. *U.S. Medicine,* July 1, 1966, p. 33.

45. R. G. Smith, "Magnesium Pemoline: Lack of Facilitation in Human Learning, Memory and Performance Tests," *Science,* 155 (1967):603; and J. T. Burns, R. F. House, F. C. Fensch, and J. G. Miller, "Effects of Magnesium Pemoline and Dextroamphetamine on Human Learning," *Science,* 155 (1967):849.

46. D. E. Cameron and L. Solyom, "Effects of Ribonucleic Acid on Memory," *Geriatrics,* 16 (1961):74–81. The so-called theoretical justification for this injection of a mixture of unidentified chemical compounds is sufficiently brief to be quoted in full:

> Theoretic indication for this treatment was the new physiological and biochemical discoveries which showed that there is a continuous process of renewal in the neuron and that RNA stimulates tissue growth. Furthermore, with the increase in yellow pigment, the RNA content of the brain decreases. To validate the theoretic assumption that increased RNA ingestion can reverse or arrest the progression of memory retention failure in the aged, oral and intravenous RNA preparations were administered to 84 patients in the last three years.

In no way do the first two sentences pertaining to tissue growth and yellow pigment even relate to memory.

MAURICE MANDELBAUM
The Johns Hopkins University

Psychology and Societal Facts

> We must, therefore, consider social phenomena in themselves as distinct from the consciously formed representations of them in the mind; we must study them objectively as things, for it is this character that they present to us.
>
> —Emile Durkheim
> *The Rules of Sociological Method*

Some years ago, in an article entitled "Societal Facts,"[1] I attempted to show that facts concerning social institutions are not reducible to facts concerning individual behavior. This amounted to a rejection of what is often termed "methodological individualism," a position made familiar through the writings of Karl Popper, J. W. N. Watkins, F. A. Hayek, Isaiah Berlin, and others. In a subsequent paper entitled "Societal Laws,"[2] I attempted to show that my position did not in fact entail those forms of holism and historical determinism which most methodological individualists have been inclined to suppose it must necessarily entail. At the end of the latter article I indicated that I would soon supplement my original discussion by attempting to clarify the relations between psychology as a discipline and the nature and existence of societal facts. It is to this problem that I at long last return.

It is all the more necessary that I do so since some of my critics have claimed that I characterized psychology in so narrow a fashion that it followed merely as a matter of definition that there are irreducible societal facts.[3] Furthermore, the term "the behavioral sciences" has in the interim become deeply entrenched; to many social scientists this term has come to suggest that psychology and the social sciences cannot remain fundamentally distinct, with each having its own types of problems. The term first became widely known through the establishment of the Center for Advanced Study in the Behavioral Sciences;[4] and while I, like other former fellows of that admirable institution, remain deeply grateful to it for aid and support, I persist in what may be my errors, still wishing to draw a distinction between the science of psychology and the systematic exploration and explanation of societal facts.

I shall lay the groundwork for this discussion by briefly considering the question of what constitutes the province of psychology. I shall then offer suggestions as to the status of societal facts, indicating the reasons why they are not to be identified with those aspects of individual behavior with which psychologists are concerned. Finally, I shall suggest that, in spite of these differences, an independent science of psychology helps to explain social institutions and some of the changes which they undergo.

I. As is well known, earlier characterizations of psychology had designated it as the science either of *mind* or of *consciousness,* but more recently it has most often been characterized as the science of *behavior.*[5] The widespread influence of the behaviorist movement has undoubtedly done much to promote such a definition, but it should be noted that a willingness to define psychology with reference to the concept of behavior is by no means confined to behaviorists. For example, in spite of their sharp opposition to the behaviorist movement, both William McDougall and Kurt Koffka chose to define psychology in terms of behavior, and McDougall had done so well before behaviorism, as a movement, had even arisen.[6] For my present purposes it will be simplest to accept such a characterization of the general province of psychology, although nothing crucial to my argument hinges on the point. What I wish first to do is inquire into the subject matter of *social* psychology, that is, into that branch, area, or field of the science of behavior which seeks to understand and explain the principles of social behavior. This question is obviously related to the topic of this essay, which is the relation of psychology to societal facts.

Although most of the issues with which social psychologists are concerned have a long history, it is not unusual to take 1908 as the date of the first attempts to transform it into a special area of investigation, for that was the year in which two texts in the field appeared: E. A. Ross's *Social Psychology* and William McDougall's *Introduction to Social Psychology.*[7] These two works adopted quite different views as to the nature of the discipline. At the time, McDougall held that all phases of human social life ultimately rest on the instinctive characteristics of men; his social psychology consisted in showing how these instinctive behavioral capacities were the foundations of the institutions characteristic of organized social life.[8] Thus, although his instinctivist approach to psychology, with its reliance upon evolutionary theory and comparative psychology, was relatively novel, his basic concerns within social psychology were wholly traditional: He wanted to discover what traits of human nature made man a social animal with capacities to create such patterns of organized life as the family, tribal organization, systems of law, religion, and so forth. The province McDougall assigned to social psychology was simply one of tracing out these connections. He assumed that once it had been shown that this was possible, it would be acknowledged that all of the other sciences which dealt with social institutions—among which he listed ethics, economics, political science, philosophy of history, sociology, and cultural anthropology—depend for their proper advancement upon the results of the science of psychology.[9]

In contrast to this approach, Ross held that social psychology had a special subject matter of its own, and that it had to proceed empirically, through directly examining each of the problems that arose within the area of its concern. In defining this area, he excluded the study of the structures of social organization, for that study he reserved for sociology. He also excluded from social psychology the behavior of individuals as individuals. Instead, Ross held that the proper province of social psychology lay in the interactions among men, and more particularly in understanding what he referred to as "the psychic planes and currents that come into existence among men in consequence of their interaction."[10] In speaking of these "planes," Ross had in mind the ways in which social interaction operates in bringing about shared actions, experience, attitudes, and beliefs; in speaking of "currents," he wished to call attention to the possibility of following patterns of change that occur within these relationships. His meaning becomes somewhat clearer when we note the key role played by *suggestibility* in his analyses. Suggestibility involves interpersonal relationships, and on the basis of suggestibility he analyzed such phenomena of interaction as crowd behavior, custom, the various ways in which beliefs and attitudes are transmitted, and so forth. All of these, on his view, were the special materials with which he thought that social psychology, as distinct from other disciplines, should deal. Thus, comparing the approaches of Ross and McDougall, it is obvious that from its very inception there were alternative views regarding the nature of social psychology, and these were connected with alternative views concerning its connection with the social sciences generally.[11]

The question of how social psychology is related to other disciplines, including other aspects of psychology, still remains an issue of major theoretical importance. Indeed, it has become even more difficult to handle because the types of problems which are now generally taken to be problems in social psychology have multiplied and at present cover a much greater range than was formerly the case. This is evident in the number of subspecialties which have developed within the general area designated as "social psychology." Nevertheless, in the period following the textbooks of Ross and McDougall, the definition of the field became fairly well standardized;[12] and in spite of recent changes in interest there has been relatively little disagreement among social psychologists as to what gives unity to their endeavors: a common concern with analyzing *the behavior of individuals in their interpersonal relationships.*[13] In other words, the subject of *social* psychology, like that of all psychology, is taken to be the behavior of individuals, but it is specifically designated as social because it is concerned with the ways in which persons interact, rather than analyzing whatever other responses an individual makes to his environment. While interactions between persons may include some factors, such as imitation, which are not as a rule present in other cases of individual behavior, the social psychologist must also be concerned with aspects of behavior with which other psychologists deal. For example, problems of perception and of the relation of perception to the nature of the stimulus, factors determining motivation, the effects of past experience on behavior, the analysis of emotional response, and so forth are of concern to psychologists generally, whether or not they are dealing with behavior that is responsive to the presence of another person. Therefore, it might seem (and has seemed to some) that social psychology is only a matter of applying the general principles of behavior to a special class of instances.

It is precisely here that important methodological questions concerning the relations between social psychology and other psychological investigations arise. In opposition to those who would hold that social-psychological questions should always be handled in terms of general principles which have been established without reference to the special data of interpersonal relationships, one *might* argue that an adequate general psychology which is applicable to human beings can never afford to neglect the factors present in interpersonal relationships. Therefore, one had better start from social psychology, applying the principles derived from these cases to all cases.[14] In defense of this alternative, it might be argued—as has sometimes been done—that insofar as human subjects are used in experiments, the responses elicited in *any* situation will be affected by socially acquired norms, and thus social-psychological concepts are presumably applicable even in experiments where no interpersonal relationships seem to be involved. Or, in opposition to either of these points of view, one might attempt to argue that the problems of social psychology should not be apprehended "from below," nor should cases in which there are no clear interpersonal relationships present be interpreted "from the top down," as if they inevitably did involve such relationships;[15] instead, it could be argued that social psychology should be regarded as a discipline emergent from general psychology, based on the principles applicable to all psychological processes but also possessing a subject matter of its own, and being capable of discovering laws of its own.[16]

I am here interested only in pointing out these possibilities. For my own present purposes the crucial question is not how social psychology may be related to other psychological investigations, but how it is related to the materials with which other social sciences, such as cultural anthropology, deal. In approaching this question it is necessary to recall that social psychology, like general psychology, is concerned with the explanation of the behavior of individuals; more specifically, it is concerned with whatever factors explain the nature of that behavior in those cases which involve interpersonal relationships. Since our lives are lived in organized societies which function only insofar as individuals act and react with respect to one another, we are constantly involved in interpersonal relationships; consequently, it is easy to suppose that social psychology will in fact constitute the single most basic science of society, as McDougall had held. Yet this is precisely the contention I wish to deny; what I wish to show is that the relations between social psychology and the other social sciences are quite different, and somewhat more complicated.

As a first step in this direction I wish to point out that even though it is natural for us to identify the field of social psychology with the behavior of individuals in organized social groups, there are other, more primitive situations in which we can observe, and in which we must seek to explain, the interacting behavior of two or more individuals. For example, there is a social psychology of animal behavior, such as the establishment of territorial boundaries in birds, the pecking order of fowl, the mating behavior of seals, or—most obviously in our own everyday life—the behavior of our domestic pets in their encounters with others of their species, and in their relations to us. All of these examples of social behavior among animals take place without reference to the organized routines of societal life.

Furthermore, in the case of humans there also are situations in which the roles and expectations which reflect the structure of a society are not the dominant factors in interpersonal relationships. For example, this is the case during the early stages of socialization in the life of the child, for the ways in which a child responds to the presence of adults (as distinct from the behavior of adults toward the child) *must* depend upon factors which cannot have been acquired initially through society, since socialization itself depends upon them. Among the factors that are important in the processes of socialization, as one can see in such studies as *The Ape and the Child* by W. N. and L. A. Kellogg, there are some which involve what might be called a differential responsiveness to *behavior;* that is, both the ape and the child responded differently to each other, and especially toward their experimenter-parents, than they did with respect to other features in the environment. Thus, forces of a specifically social-psychological nature are operative in early behavior, and it is largely on the basis of them that training proceeds. However, these traits do not simply disappear in adult life: We respond differently according to the ways in which others respond to us, and traits such as unconscious imitation, which are clearly important in the socialization of the child, are present in adult behavior as well. Therefore, it should not be assumed that, in investigating interpersonal behavior, the social psychologist is necessarily studying relationships for which the institutional structures of a society are responsible; in some cases at least, he is dealing with a different stratum of behavior. This difference can perhaps be expressed by saying that the social psychologist investigates behavior from the point of view of how responsive interaction takes place among *individuals,* rather than investigating the sociological question of how various *roles* are related in the organized life of a society.

This point can also be made with respect to phenomena such as leadership and group formation. While, under normal circumstances, the ways in which persons interact is circumscribed and channeled by socially ascribed roles and resulting expectations, it is possible under experimental conditions to reduce the effect of these factors to a minimum. This has, for example, been done in psychodynamics experiments such as those conducted at Bethel, Maine, where leadership and group formation were studied in what, from a sociological point of view, must be called "empty environments." Because these environments were sociologically unstructured, it would be a mistake to expect (as sometimes seemed to be expected) that the results of such experiments could be directly transposed into the context of everyday life, since in everyday life socially ascribed roles *do* profoundly influence behavior. However, it would be surprising if the factors influencing leadership and group formation under these special conditions did not have an influence upon the dynamics of groups when sociological factors are also present. In fact, there seems to be sufficient evidence drawn from other studies in group dynamics to say that this natural expectation is often fulfilled. What may seem strange is that I should be using this example, and that of the socialization of the child, as my first step in attempting to show that social psychology is *not* to be considered as the single most basic science of society, since in both examples I have tried to make clear that social psychology investigates factors in interpersonal behavior which do not themselves derive from facts concerning social organization. That my argument is not odd can quickly be made clear.

In showing that there are factors determining interpersonal behavior which are *not* sociologically generated variables, and that these factors are presumably important in understanding personal interactions wherever they are found, I am indicating that the province of social psychology, as an explanatory discipline, does not cover the whole range of those factors to which one must appeal in attempting to explain why, in organized societies, interpersonal relations take the forms that they do. Social psychology does not cover the whole range of these factors precisely because its principles are relevant to *all* interpersonal relationships, including those in which sociological factors fail to play any role. To explain behavior in which the latter factors *do* play a role demands the introduction of concepts which fall outside the scope of social-psychological analyses. As we shall see, these factors are not themselves instances of interpersonal behavior: They serve to structure such behavior without being identical with it. Showing that it is not arbitrary to suppose that this is the case will constitute the next step in the present argument.

II. It should now be familiar, through the writings of Ludwig Wittgenstein and others, that rules and conventions may serve to control human actions without our identifying these rules with the specific behavioral acts they govern. The knight's move in chess and what makes up a home run in baseball are components of the structure of particular games. While the acts performed by individual players have reference to the rules that are definitive of these games, the rules cannot be identified with the performances of those whose behavior they regulate. I shall soon use this as an analogy in order to help clarify my own position. In the meantime, however, I wish to take note of the fact that some social theorists have used the notion of rule-governed behavior not merely as *an analogy* that can be helpful in characterizing the nature of social organization; instead, they have *identified* the structure of a society with the rule-governed behavior of the individuals within that society.[17] This is an equation I wish to reject.

It is, of course, easy to see that there are many ways in which the functioning of a society is dependent upon the fact that individuals do behave according to common rules. Unless they followed rules, and unless much of their behavior were governed by these rules, no consistent pattern of interaction would come into existence, and there would be no societies. In this connection we may note that, as others have pointed out, a society is not like an object such as an automobile, which is made up of parts that exist whether or not they are functioning. The existence of a society is inseparable from the ongoing functions that individuals perform.[18] Let us refer to the common ways of doing things as *practices,* stressing the fact that a practice is simply a widely shared way in which the individuals of a particular society behave. Let us also note that the consistency and the stability of many practices depend upon the fact that these practices are governed by specific rules that people learn, and that they accept. On such a view, a language can be regarded as a rule-governed practice, as can other socially acquired skills, such as gardening, basket-weaving, canoe-making, healing, and the like. Given the prevalence of such rule-governed ways of behaving, it might seem as if one could describe any society in terms of the particular system of practices engaged in by those who live in that society. On such a description, sociologists and anthropologists would be expected to analyze the nature of a society, and of any changes occurring within it, in terms of the interrelated rule-guided activities of individuals.[19] However, that is precisely the view I wish to reject, and I shall now briefly point out certain difficulties in it.

In the first place, and most obviously, there are some factors which are important in analyzing the structure and functioning of a society which are not themselves instances of rule-guided behavior. For example, any anthropologist must take into account the size and density of a population, and the resources in land and water available to those inhabiting a particular territory. To be sure, factors of this kind may themselves be affected by rule-governed practices, as size of population is affected by female infanticide, or as irrigation alters the availability of arable land. However, regardless of the ways in which these factors are to be explained, the actual size and density of a population, and the resources available at any time, are not themselves *practices*. Nonetheless, they do affect practices—including, perhaps, the fact that a society begins to practice, and continues to practice, female infanticide, or employs irrigation to develop the land. Therefore, if one attempted to analyze societies in terms of rule-governed interactions of individuals *only*, what is characteristic in these practices, and the changes which they undergo, would in many cases (and perhaps in most cases) be left unexplained.

This situation obtains not merely with respect to factors such as size of population or available resources, which might in some sense be regarded as nonsocial, "physical" factors; it applies in other instances as well. For example, a tribe may come into contact with a neighboring tribe, and may develop magical practices and warlike practices that it takes to be essential to its own defense. In such cases, what is involved is the initiation of a new set of internal practices because of fear of some external factor, and the new practices based on such fears may then affect other practices of the tribe.[20] Thus, just as the soil and the climate, or the size and the density of a population, affect various features of the life of a society, so too may the presence of neighboring societies. Therefore, in accounting for a particular set of practices and the changes which various practices undergo, one cannot simply describe the ways in which the individuals within a society interact with one another: One source of their practices will be found in beliefs, expectations, hopes, and fears connected with the environment in which they live. The importance of this fact becomes particularly evident when we note that the external environment is not always regarded in a merely naturalistic way; in many cases it is viewed as including all manner of supernatural powers or forces. Thus, if we are to understand the actual practices of a society—that is, the ordered ways in which individuals interact with one another in obtaining their food, constructing their shelters, ensuring their crops, and warding off their enemies—we must take into account what, for want of a better term, I shall call the *representations* they share.[21]

While the shared representations to which I have thus far called attention have been beliefs, expectations, hopes, and fears that were specifically connected with various aspects of the environment, there also are other types of shared representations that help to account for the behavior of the individuals within a society. For example, there is an awareness of the expectations of others, and a recognition of the sanctions that may be imposed for certain types of action. Only to the extent that such expectations are mutually recognized and widely shared are they effective in creating stable sets of relations within a society. The forms of behavior to which they give rise are those which, following present usage, I have designated as *practices.* What I find it important to recognize—and what is often not recognized—is that a distinction is to be drawn between the shared representations which govern practices and these practices themselves. As ethnologists have shown, over and over again, if we are to understand a set of practices we must understand the beliefs, expectations, and other shared representations these practices involve. That this is true may be suggested by considering the situation in which one finds oneself when watching an unfamiliar game being played. In order to understand the behavior of those who are playing, one must come to understand the rules of that game, since it is with reference to these rules that their behavior is governed. It is at precisely this point that I find it helpful to draw an analogy between the rules of a game and the structural features of a society.

From the point of view of a participant, the rules of a game are not to be identified with his own behavior, or with the behavior of those who participate with him in that game: The rules lay down principles to which the behavior of each of the participants is to conform. Thus, a set of rules is not identifiable with a set of behavioral acts, but has a different status. Such rules are what I have called shared representations. It is in terms of them that one participant can claim a foul against another, and can argue whether the action of his opponent did or did not conform to mutually accepted rules. This situation parallels what is to be found in the laws of a society, whether these laws are customary or codified. Whatever the school of the legal realists may have contended, laws are not summary descriptions of the behavior patterns of any set of individuals; they are prescriptions as to how individuals—including judges and policemen—are to behave.[22] Even in those areas in which no established laws or rules exist and no well-entrenched customs have developed, the interaction of persons engaged in common pursuits depends upon their mutual expectations: These expectations govern their behavior and are not to be identified with it. Thus, throughout any society, the interactions of persons engaging in common practices depend upon what I have termed shared representations.

The fact that all expectations, beliefs, hopes, fears, and so forth are always the expectations, beliefs, hopes, or fears of *someone* entails the fact that even though these representations are shared, they are shared by individuals. Consequently, it is all too readily assumed that such representations are in some sense "internal" or "mental"; in other words, that they are to be construed merely as facts about those who recognize them. It is therefore commonly assumed that if one characterizes rules, or other shared representations, as "objective," one is treating that which exists only in the minds of individuals as if it existed elsewhere. The *being* of a rule, it might be said, is merely its being recognized as a rule. However, if this is taken as meaning that a rule is a fact about the individuals who recognize that rule, it is a badly mistaken supposition.

Consider, for example, an analogous case in which a person performs some simple arithmetical operation, such as multiplying two three-digit numbers. This is a mental operation involving a set of rules, but the rules are not themselves to be taken as if they were facts concerning the person who performs the multiplication; rather, they are recognized by him as having a status that is independent of him, and it is with reference to these rules that he performs the operations he does perform. To be concerned with facts about the particular person who performs the multiplication would be to ask, for example, whether he performed it rapidly or slowly as compared with other persons; or it would be to ask at what age, and under what condition, he learned to multiply numbers, or to multiply numbers of this sort. It is also possible to be interested in psychological facts that are more general, such as how we are to account for differences in the abilities of children to multiply numbers, or what principles of learning are applicable to the teaching of multiplication in schools. Also, one may be interested in even more general psychological questions which concern the ability of human beings to deal with abstract concepts, to remember and apply rules, and the like. However, in these cases the rules—and their acceptance—are taken as given: The problem is not one of understanding the rules themselves, but of understanding something about individuals, and about their ability to use these rules. Thus, remaining within the context of this example, we may say that psychology, as a science of human behavior, presumably has a great deal to say about such questions as how individuals learn mathematics; however, rules such as those of multiplication cannot in any sense be considered to be psychological principles or laws.

Similarly, it is my contention that psychology, although it deals with the principles or laws of how individuals behave under varying conditions, is not a science which has as its task the description or the analysis of those conditions. They are taken as *given.* Among the important given conditions that must be taken into account by psychology is the fact that human individuals are born into societies, are reared in societies, and that in these societies they are forced to learn the rules which govern interpersonal relations. It is my contention that, on the analogy of the case of mathematics, psychology can explain a great deal about how individuals learn rules, how they react differentially to rules, and the like. What psychology does *not* explain is the existence and the nature of those rules which govern the behavior of socialized individuals.

While it might be acknowledged that this contention is true with respect to other fields of psychology, some might wish to claim that it is not true with respect to social psychology, that the task which social psychology sets before itself is that of explaining the rules to be found in social interaction. In order to conclude my argument I must suggest why this is not the case, and why, therefore, my contention also applies to social psychology.

Let me begin by admitting what is altogether obvious, that shared representations, such as rules of behavior prescribed in various types of situation, presuppose interaction among persons: Without such interaction, rules would not come into existence, nor would they be learned. Thus, whatever general principles or laws apply with respect to interpersonal behavior will be of possible relevance to an explanation of how (through the interaction of persons) rules come into existence. For example, any general principles which can be formulated concerning relations of dominance and submission might be relevant here. Furthermore, social psychology may well be relevant to various aspects of the processes through which rules are initially learned, or how they are reinforced, or under what conditions their authority over the individual tends to increase or to diminish. However, one must immediately note that any principles that would be effective in answering such questions could not be applicable to one set of rules only, or to one society only, but would have to be of general import, applying under diverse conditions. It is at this point that one can see that social psychology will not provide an explanation of the host of diverse rules which characterize different societies, nor will it explain the subsets of the rules within a society which apply to persons who occupy different stations and who therefore function in different roles.

This limitation of social psychology as an explanation of societal facts may perhaps be made most clear by pointing out that when a psychologist seeks to account for the ways in which individuals behave with respect to one another in a particular society, he must understand the structure and the operative rules of that society, and must know how—in societal terms—the individuals in question stand in relation to one another. To be sure, as I have already remarked, in some interpersonal relationships societal roles may not play a determining part, or may play only a minor part. However, for the social psychologist who is dealing with human and not animal subjects, societal facts must always be acknowledged to be among the *possibly relevant* conditions that are to be taken into account. In other words, societal facts may always prove to be among the initial conditions which it is necessary to know if one is to explain the behavior of an individual. As relevant initial conditions, they are descriptive data to which the general principles, or laws, of social psychologists are to be applied. When this is recognized one can see that it is no more the task of empirically minded social psychologists to explain these initial conditions than it is the task of a physicist to explain the existence of a particular set of initial conditions when he applies, say, the laws of mechanics to this set of conditions and deduces what will then follow.[23] The description of what constitutes the relevant initial conditions in cases of interpersonal behavior is not in most instances easy. However, when one recognizes that the situations faced by individuals in most of their relationships with others are partly structured by common expectations and by rules, the task of description is far less hopeless than would be the case if the only determining conditions of individual behavior were facts about the individual himself. Thus, the explanatory task of the social psychologist is simplified because societal facts help structure the individual's world.

If what has been said is true, there is no reason to suppose that social psychology can supplant ethnology and sociology as *the* basic discipline in understanding the behavior of human beings in their interpersonal relationships. However, there is also no reason to suppose that, on the contrary, sociologists or ethnologists can take over the social-psychological task of searching for general laws which adequately explain the behavior of individuals in their interpersonal relations. Nor do I see any reason to believe that a new interdisciplinary behavioral science is called for, in which culture and personality become hyphenated, and social psychology, cultural anthropology, and group-behavior theory become a unified science. In what follows I shall suggest a different way in which a social-psychological approach and a societal approach to human behavior may supplement each other without losing their separate identities.

III. In my article "Societal Laws," to which I have already referred, I argued that the attempt to establish laws concerning social institutions need not involve any objectionable form of holism or of historical determinism. I shall not repeat that argument here. However, it is relevant to suggest—although only briefly—how the existence of shared representations, such as rules, makes it plausible to assume that there are the sorts of societal laws which, in that article, I wished to defend. On the basis of what I shall say in this connection, it will also become clear that I attach more importance to the explanatory power of psychological laws than my emphasis up to this point may have led one to suppose.

In speaking of societal laws I here wish to confine myself to the particular type of law which, in the article cited, I was interested in defending. That type of law would attempt to show that a particular institution is always present with, and varies with, some other institution, or that a similar relationship obtains with respect to two or more aspects of a particular type of institution. Thus, I am *not* here concerned with whether or not there are general patterns or laws in history, but with the sort of law with which, for example, E. B. Tylor was concerned when he attempted to show with respect to kinship and marriage that there was a covariance between rules of residence and rules of descent.[24] I am not, of course, attempting to offer an explanation of the particular covariance that Tylor believed one could establish, nor am I defending the claim that he had established such a covariance. I am only interested in showing that—if what I have said up to this point is true—it is not surprising that there should be connections among specific sets of societal facts, and that these connections might be formulable as societal laws.

Let us suppose, as I have argued, that behind any particular practice there do exist shared representations in the form of common beliefs, expectations, and the like. While a variety of different practices might be compatible with the specific nature of these representations, it would surely be true that other practices might not be compatible with them. Such practices would therefore not develop except in communities where other, quite different representations were to be found. On the basis of this assumption, one would be in a position to hold that even when two similar sets of shared representations were dependent upon very different originating circumstances, their consequences would be similar, and generalizations relating particular practices to specific kinds of shared representations might be found. In such cases, the historical question of how these representations might have arisen would be irrelevant to the formulation of a law, just as in the natural sciences it is possible to apply a law describing a functional relationship without first accounting for the set of conditions to which that law applies.[25] This, I believe, lends weight to the functionalist position in anthropology, and to functionalist criticisms of a historical approach.[26] Furthermore, the possibility of establishing laws concerning the practices of different societies is not confined to showing that only certain sets of practices are compatible with certain types of shared representations; it is also the case that there may be direct relations of compatibility and incompatibility between various practices themselves. For example, if a specific form of agriculture is practiced in a community, the fact that a set of individuals engage in this practice may make it literally impossible for them to engage in various other activities, thus entrenching a particular division of labor in that society. In many cases such a division of labor will be reciprocal, so that if those who engage in one practice, *a,* cannot engage in some other practice, *b,* and those who engage in *b* may not also be able to engage in *a.* In such cases it would be reasonable to speak of *co-related practices.* In the case of such co-related practices it is obvious that a change in either practice, owing to changes in technology, in the size of population, or in any other factors, would in all likelihood affect the co-related practice. And since any one practice may be co-related to a number of different practices, the repercussions of a change in one may be felt in many aspects of the life of that society.

The type of hypothetical case I have been describing may justifiably be said to be very coarse-grained. Very possibly it would only be approximated in small communities with a relatively simple division of labor. However, the same point concerning the mutual compatibility or incompatibility of specific practices would be found in, say, a sophisticated monetary economy where the supply of money at any one time has repercussions on the practices characteristic of various segments of the population who borrow and lend money for a variety of different purposes. Laws concerning the relations among the variables within an economic system are (I suggest) examples of how co-related practices affect one another when a particular set of initial conditions is given and the relevant boundary conditions are assumed to remain stable. Thus, it need not be supposed that the only type of societal laws which one may expect to find are those which concern the ways in which particular types of practices may be related to particular types of shared representations; it is also possible that various practices are co-related in ways which make it feasible to trace functional relationships among *them.* However, it is not my present task to develop this point any further. What I now wish to suggest in concluding this discussion of "psychology and societal facts" is that the existence of such specifically societal laws is not in any way incompatible with the fact that there may *also* be social-psychological laws which are of importance in understanding social organization, and in explaining whatever changes a society may undergo.

In this connection it is first important to note that there is nothing surprising in holding that any given set of circumstances may demand that we use multiple sets of laws in attempting to offer an explanation of it. For example, in explaining the action of an internal combustion engine, one must invoke laws concerning the expansion of gases when ignited, the laws of the lever, of the conditions under which friction is reduced, and the like. One need not reduce these various laws to one another, nor to any one set of fundamental and all-encompassing laws; they can, so to speak, be intersecting laws which together explain the events that occur. Thus, it should not be surprising that in offering an account of the nature and the changes in any society one might have to invoke both societal and social-psychological laws. In fact, it would be surprising were this not the case. As we have seen, although it is a mistake to equate the shared representations according to which individuals govern their behavior with that behavior itself, still we are dealing with individuals who are aware of one another, and who interact with one another in carrying out the various practices to which these representations give rise. Therefore, whatever general principles of interpersonal behavior social psychologists can establish will be applicable when there is an interaction among persons to carry on the practices characteristic of a particular society, or of some differentiated group within that society.

Nor should we suppose that it is only social psychology which is relevant to an understanding of the behavior of individuals within the matrix provided by society. Questions concerning motivation, or learning, or the development of personality, can be expected to be relevant to understanding how a particular form of societal structure will affect the behavior of individuals. Bearing this in mind, it is not enough to insist, as I have been insisting, that even a completed science of psychology would not be sufficient to enable us to understand the structure of societies; it is no less important to insist that if we are to understand how the practices of a society are maintained we must also take general psychological principles into account, not merely seeking to discover co-relations among societal facts. These arguments lead, then, to a single conclusion: To gain any measure of concrete understanding of the nature and the changes occurring in any society, we must apply intersecting sets of laws to existing societal structures, seeking to delimit the psychological possibilities for action under those circumstances, and establishing the psychological processes which may be presumed to be operative under such conditions. On the other hand, we must also recognize that psychological factors only operate under specific sets of circumstances, and the shared representations underlying societal life at any one time and place will provide a set of constraining conditions which those who belong to a particular society find themselves forced to take into account. Thus, psychological inquiry and the analysis of connections among societal facts proceed independently, but they must intersect whenever we seek to explain the concrete forces that are present within any society. Were one to seek a more all-embracing set of laws, more general than the formulations to be found in psychology or in the analysis of societal facts, one would be attempting to create that all-embracing social science which includes so much that it fails to allow us to distinguish among the various factors each of which, in its own special way, helps to explain those complex phenomena which the social sciences, taken as a whole, seek to explain.

NOTES

This paper was presented as a lecture at the University of Pittsburgh in March 1970. No attempt has been made to relate it to other, more recent discussions of the problem.

1. *British Journal of Sociology,* 6 (1955):305–17. This article is also to be found in Patrick Gardiner, ed., *Theories of History* (Glencoe, Ill.: The Free Press, 1959), pp. 476–88; in E. H. Madden, ed., *The Structure of Science* (Boston: Houghton Mifflin, 1960), pp. 166–76; and in *Bobbs-Merrill Reprint Series in Philosophy,* Phil. 128 (Indianapolis, n.d.).

2. *British Journal for the Philosophy of Science,* 8 (1957):211–24. This article is also to be found in W. H. Dray, ed., *Philosophical Analysis and History* (New York, Harper and Row, 1966), pp. 330–46.

3. See J. W. N. Watkins, "Historical Explanation in the Social Sciences," *British Journal for the Philosophy of Science,* 8 (1957):108 n.; and Alan Donagan, "Social Science and Historical Antinomianism," *Revue Internationale de Philosophie,* 11 (1957):444–45.

4. Although there were a few earlier uses of the term "the behavioral sciences," it undoubtedly came into prominence with the founding of the center, which was sponsored by the Ford Foundation. According to Preston Cutler, the present associate director of the center, the term was used with reference to the future work of the center by D. Marquis and J. G. Miller during the planning conferences which led to its establishment. (Concerning immediately prior uses of the term at the University of Chicago, see J. G. Miller in *American Psychologist,* 10 [1955]:513–14. It is also to be noted that in *Principles of Behavior* [New York: Appleton-Century-Crofts, 1943] Clark Hull spoke of "the behavior sciences" and regarded it as possible to construct a systematic theory of behavior which would be foundational for "social" sciences [pp. 17 and 398–401].)

In the present announcements of the center, which suggest to fellowship candidates the areas of advanced study which are appropriate, no definition of "the behavioral sciences" is offered; there is simply a list of a wide variety of disciplines, such as anthropology, education, psychology, economics, philosophy, and so forth. However, in two earlier documents (the *Report of the Behavioral Science Division of the Ford Foundation* of June 1953, and a staff memorandum dated June 5, 1954, written by Ralph W. Tyler, the first director of the center), it is clear that a distinction was intended to be drawn between the behavioral sciences and the social sciences. Neither document attempted to state the distinction in rigorous fashion; each relied upon a few illustrative examples to suggest its nature. The two sets of examples are not incompatible, but they have little in common; consequently, their joint use is not particularly helpful in clarifying the original meaning which was to attach to the term "the behavioral sciences."

The fullest account of the introduction of the term, with an indication of its widespread adoption in recent literature, is to be found in B. Berelson's article "Behavioral Science," in the *International Encyclopedia of the Social Sciences.*

5. For an extended discussion of the province of the science of psychology, see my article "To What Does the Term 'Psychology' Refer?" in *Studies in History and Philosophy of Science,* 2 (1972):347–61.

6. In 1905, in his *Physiological Psychology* (London: Dent, 1905), McDougall defined

psychology as "the positive science of the conduct of living creatures" (p. 1); in 1912, he contributed a volume to the Home University Library series entitled *Psychology: The Study of Behavior* (New York, Holt).

In *The Growth of the Mind* (New York: Harcourt Brace, 1925), Koffka defined psychology as "the scientific study of living creatures in their contact with the outer world" (p. 4); in the *Principles of Gestalt Psychology* (New York: Harcourt Brace, 1935), he said:

> Although psychology was reared as the science of consciousness or mind, we shall choose behavior as our keystone. That does not mean that I regard the old definitions as completely wrong—it would be strange if a science had developed on entirely wrong assumptions—but it means that if we start with behavior it is easier to find a place for consciousness and mind than it is to find a place for behavior if we start with mind or consciousness. (p. 25)

7. As John Dewey pointed out in "The Need for Social Psychology" (*Psychological Review,* 24 [1917]:266 ff.), the year 1890 might also have been chosen, for in that year William James published his *Principles of Psychology,* which called attention to some important social-psychological problems, and Gabriel Tarde published his influential work, *Les lois de l'imitation.*

8. Later, in *The Group Mind* (1920), McDougall took a quite different approach which was nearer that of Ross. He had planned a further volume, which might have established the connections between the first two, but that volume was never written; as matters stand, his two contributions to the field are not contradictory, but they proceed on the basis of wholly different assumptions. (On this and other problems in the history of social psychology during the nineteenth century and the first two decades of the twentieth, see F. B. Karpf, *American Social Psychology* [New York: McGraw-Hill, 1932].)

9. *Introduction to Social Psychology,* rev. ed. (Boston: Luce, 1926), pp. 1–19.

10. *Social Psychology* (New York: Macmillan, 1908), p. 1.

11. Still another view was, of course, characteristic of Auguste Comte's position. Psychology did not exist as a separate discipline, according to him: Individual psychology was a branch of biology, whereas social psychology was included within sociology.

12. For a tabulation of definitions of social psychology drawn from twenty-two textbooks published between 1908 and 1934, see H. Cantril, "The Social Psychology of Everyday Life," *Psychological Bulletin,* 31 (1934):297–330. His extensive use of references provides an interesting bibliography of over three hundred items published within the same period.

13. My remarks are based chiefly, but not exclusively, upon the following: David Krech and Richard Crutchfield, *Theory and Problems of Social Psychology* (New York: McGraw Hill, 1948); S. E. Asch, *Social Psychology* (New York: Prentice-Hall, 1952); Muzafer Sherif and Carolyn W. Sherif, *Social Psychology,* 2d ed. (New York: Harper and Row, 1956); David Krech, Richard Crutchfield, and E. L. Ballachey, *Individual in Society* (New York: McGraw Hill, 1962); E. E. Sampson, *Approaches, Contexts, and Problems in Psychology* (Englewood Cliffs, N.J.: Prentice-Hall, 1964); and E. E. Jones and H. B. Herard, *Foundations of Social Psychology* (New York: Wiley, 1967).

As one among several exceptions to my generalization regarding the definition of social psychology in terms of the behavior of individuals I should especially call attention to Daniel Katz and Robert L. Kahn, who, in the introductory chapter of their *Social Psychology of Organizations* (New York: Wiley, 1966), have explicitly taken note of the methodological issues involved. Their attempt to apply social-psychological concepts to the behavior of organizations, rather than to individuals only, resembles the approach of McDougall in *The Group Mind;* however, unlike McDougall, they avoid any semblance of speaking of organizations as if such organizations possessed mental traits.

As another, and earlier, exception the position of J. F. Brown should be noted. Although he defined the province of social psychology as being the investigation of "the behavior and reactions of the individual with regard to his fellow men, whether as other individuals or as groups" (*Psychology and the Social Order* [New York: McGraw-Hill, 1936], p. 3), and thus seems to be concerned only with factors influencing individual behavior, his actual procedure led him to apply his explanatory concepts, not only to social institutions, but to individuals as well.

14. This was the position apparently advocated by Dewey, in expressing the hope that the development of social psychology would exert influence on general psychology (*Psychological Review,* 24 [1917]:271–72).

The position of George Herbert Mead also deserves mention here. While he admitted the independence of physiological psychology from social psychology, he regarded these two disciplines as parallel inquiries; all nonphysiological inquiries into mind or self he took to be dependent upon the basic principles of social psychology. Among his discussions of this topic, two may be singled out for attention: "Social Psychology as Counterpart to Physiological Psychology," *Psychological Bulletin,* 6 (1909):401–08; and *Mind, Self, and Society* (Chicago: University of Chicago Press, 1934), pp. 222–26.

15. I borrow these phrases from the introductory chapter of Gustav Fechner's *Vorschule der Aesthetik* (Leipzig: Breitkopf and Härtel, 1876).

16. A position of this general stamp would seem to be adopted by S. E. Asch in his *Social Psychology.* Particularly relevant are his first and ninth chapters.

17. The most extreme example of this seems to me to be the position adopted by Peter Winch in *The Idea of a Social Science* (London: Routledge and Kegan Paul, 1958), in which he conceives of all social relations as being expressions of the ideas men hold regarding reality (see, for example, pp. 23–24). His conception of the philosophy of science makes it unnecessary for him to argue for that interpretation through any appeals to matters of empirical fact (see pp. 17–18 and 20); yet it is nonetheless strange that he fails to inquire what can be supposed to shape men's ideas of reality in such a way that, at different times and places, different "forms of life" arise.

18. I borrow this contrast from Daniel Katz and Robert L. Kahn, who phrase it as follows:

> Physical or biological systems such as automobiles or organisms have anatomical structures which can be identified even when they are not functioning. In other words, these systems have an anatomy and a physiology. There is no anatomy to a social system in this sense. When a biological organism ceases to function, the physical body is still present and its anatomy can be examined in a post-mortem analysis. When a social system ceases to function, there is no longer an identifiable structure. (*Social Psychology of Organizations,* p. 31)

19. If I am not mistaken, this would be an implication of the view held by Alan Ross Anderson and Omar Khayyam Moore in "Toward a Formal Analysis of Cultural Objects," *Synthese,* 14 (1962):144–70. After surveying alternative definitions of culture, they conclude that cultural objects should be defined as consisting of all, and only, things which are *learnable.* As they point out (pp. 165–66), such a definition focuses attention on such things as "propositions, techniques, values, rules, and the like," rather than on institutional structures, or other *products* which result from the fact that people act in the ways in which they have learned to act.

An analogous position may perhaps be attributed to Melville J. Herskovits on the basis of his discussion of "Culture and Society" in *Man and His Works* (New York: Knopf, 1947), chap. 3. It also seems to be implicit in David Bidney's view that it is only for heuristic purposes that facts of interest to anthropologists and sociologists can be treated apart from the activities of the human organisms on which they depend (*Theoretical Anthropology* [New York: Columbia, 1953], pp. 48–49, 106–07 et passim).

20. Concrete evidence for the internal relatedness of the various practices of a society was one of the major contributions of early functionalist theory in anthropology. On the history of functionalist views, see my "Functionalism in Social Anthropology," in *Philosophy, Science, and Method: Essays in Honor of Ernest Nagel,* ed. Sidney Morgenbesser et al. (New York: St. Martin's Press, 1969), pp. 306–32.

21. It will be recognized that my use of this term has some degree of affinity with Emile Durkheim's concept of "collective representations," and I gladly acknowledge that affinity. However, as will become clear, the unfortunate assumption of some sort of "collective mind" which has frequently (and with some justice) been attributed to Durkheim's view of "collective representations" has no place in my use of the term.

I also wish to point out that I am using "representations" as a technical term and am not borrowing from the notion of representation as that term has been used in discussions of art, nor as it has sometimes been used in epistemological theories.

22. The question of what brings about changes in custom or in laws, and in the interpretation and enforcement of laws, is not my present concern. These are the areas in which the views of legal realists may appear most convincing. However, it should be noticed that it is with reference to the behavior of *judges,* and of enforcement agencies, that they define the law, not with respect to the behavior of other persons in the society. Thus, from the point of view of the members of a society (including judges and policemen), behavior is expected to conform to the law, and the law is not taken to be identical with the behavior it governs.

23. To be sure, both the physicist and the social psychologist may formulate for themselves the problem of explaining how a particular set of initial conditions came to occur. They may do so in terms of general laws. However, in *that* explanation they must assume another, anterior set of initial conditions which it is not their purpose to explain.

24. Tylor, "On a Method of Investigating the Development of Institutions: Applied to Marriage and Descent," *Journal of the Royal Anthropological Institute,* 18 (1889):245–69.

25. For a discussion of what I regard as functional laws I should like to refer the reader not only to my article, "Societal Facts," which has already been cited, but to my book *History, Man, and Reason* (Baltimore: Johns Hopkins Press, 1971), pp. 114–27.

26. I regret having failed to notice this possibility when I wrote the article "Functionalism in Social Anthropology," which I have already cited. While my present suggestion does not alter the particular criticisms which I leveled against the functionalist school, it would have altered the tone of my conclusions by showing that there are other respects in which the functionalist thesis can be of positive use.

It may also be that the assumptions I am making here concerning the relation between representations and practices, and the relations among practices that I shall immediately discuss, throw some light on the phenomenon of "convergence." For a brief discussion of this concept, see Alfred Kroeber, *Anthropology* (New York: Harcourt Brace, 1948), pp. 539–41.

Index of Names

Index of Topics